Advance Praise for *Disrupting Program Evaluation and Mixed Methods Research for a More Just Society: The Contributions of Jennifer C. Greene*

A brilliant depiction of Jennifer's career. The reflections of colleagues, mentees, and friends take us on a five-decade journey of one of the evaluation field's most prolific scholars and practitioners. *Disrupting Program Evaluation and Mixed Methods Research for a More Just Society: The Contributions of Jennifer C. Greene* is a beautiful tapestry of her career and enduring legacy. It is an intimate exploration of the convictions that made her a disruptor and her unwavering commitment to the values of democracy, equity, and justice. This book is not only about Jennifer as the consummate professional, it also tells a story about relationships that have been forged with love and compassion.

—**Maurice Samuels**, PhD
Managing Director of Evaluation and Learning, Sierra Health Foundation

I have been a longtime friend and colleague of Jennifer Greene and have admired her work for several decades now. On numerous occasions her insight and wisdom has helped to shape my own thinking and contributions. I am absolutely delighted to see this volume coming out. Such an amazing testament to her many faceted substantive and methodological contributions to our glorious field over the trajectory of her career. The list of contributors to this volume is impressive. These perspectives nicely span the range of Jennifer's many and varied contributions and provide exceptional insight into her thinking and the impact of her work. The editors' choice to give Jennifer "the last word" is a very good one, and the final interview with her, very compelling reading. This volume should find its way onto the bookshelf of anyone interested in understanding diversity in our field and looking for a coherent vision of what the field can accomplish.

—**J. Bradley Cousins**
Professor Emeritus of Evaluation at the Faculty of Education, University of Ottawa

Have you read the list of contributors who have written about the contribution of Jennifer C. Greene to research and evaluation? Their participation speaks volumes to the wide-reaching influence that Jennifer has had as she has stirred, provoked and challenged us to be better practitioners and better people. Because of her body of work there are no excuses for us not visiting with people, listening deeply to them, and applying our skills to making the world a better place for them to live in. The contributors speak about this and so much more, with the volume providing us with an excellent opportunity to learn from those who are part of Jennifer's succession plan.

—**Fiona Cram**
Katoa Ltd, Aotearoa New Zealand

This book is a gift to all of us who cherish Jennifer as a colleague and friend and to those future generations of evaluators who will be guided by her rich legacy. Jennifer is a consummate professional, a thoughtful practitioner, a versatile theorist, a generous mentor, and a conscientious citizen for justice. The editors of this volume have skillfully curated a balanced collection of authors from within and beyond the academy, emerging scholars and established elders, bringing both domestic and international perspectives. Collectively, they map the breadth of Jennifer's influence and the depth of her scholarship. The closing chapter joyfully captures Jennifer's wit, humility, convictions, kindness, and caring in her own words.

—**Karen E. Kirkhart**
Professor Emerita, Syracuse University

Jennifer Greene has had an indubitable impact on the field of evaluation and particularly culturally and contextually relevant and responsive evaluation. During graduate school, her courses exposed me to an entire field for which I had no knowledge, but that resonated intuitively. Her focus on dialogue, on decentering often privileged voices aligned with a desire to dig deeper in evaluation methods to go beyond the surface or the obvious. Her commitment to the values of democracy, diversity, and dialogue transformed my way of knowing and of understanding and set me and my colleagues up for greater impact in the evaluation field.

Professor Greene's commitment to diversity shone brightly in the diversity of her students from so many backgrounds. Working in public policy and advocacy, Professor Greene's emphasis on inclusive, culturally responsive, and educative theories, methods, and pedagogical practices has been extraordinarily useful for my work in public policy evaluation both in the United States and internationally. For those of us whose lives are committed to endeavors that will bring about a more just world, this book is invaluable.

—**Amara C. Enyia** JD, PhD
President, Global Black, Manager for Policy & Research, Movement for Black Lives

By describing and reflecting on Jennifer Greene's considerable contributions to the theory and practice of evaluation, this book both honors her legacy and ensures that it will continue to influence and inspire the field. The chapters show how Jennifer's "disruptions" start with a thorough understanding of what came before and include ideas for paths forward. While the book's greatest value is in the descriptions of Jennifer's scholarship and influence on the field, the book is enhanced by the authors' perspectives on Jennifer as a teacher, mentor, social justice advocate, scholar, friend, and colleague. Readers will be inspired to be as thoughtful, intentional, and as clear as Jennifer is in those roles and others.

—**Leslie J. Cooksy**
Independent Consultant and 2010 AEA President

This book is a wonderful documentation of the contributions of Jennifer Greene to advancing evaluation and mixed methods research! One might read this compilation as different points of view on Jennifer's innovations and disruptions, maybe even imagining a dialogue among the authors that weaves together the ways Jennifer's work has been influential. But this book is also a testimony to Jennifer the mentor, friend, leader, and influencer; while Jennifer was disrupting what it meant to conduct evaluation, she was also creating space for colleagues to explore and invent and new voices to claim their own approaches to evaluation. This book represents both her influence on ideas and her influence on evaluation practitioners, scholars, policymakers, students, colleagues, and friends. And that influence, as demonstrated in this volume, has helped to make evaluation more dynamic, humanistic, and justice-oriented.

—**Leslie Goodyear**
Principal Evaluation Director, Education Development Center

This book, a Festschrift for Jennifer Greene, provides a powerful tribute to a pioneering spirit in our evaluation field. Jennifer has tirelessly pushed for democratization in evaluation, the inclusion of diverse perspectives, and the need to continually address and challenge power imbalances. Her trailblazing work in mixed methods evaluation continues to educate and influence evaluators around the globe. Her contributions to evaluation have been foundational, not only through her writing but her mentorship of many and her deliberate efforts to ensure that our evaluation community itself is strengthened by the talents of a diverse set of scholar-practitioners. The voices of those she has helped to raise are prominent in this volume and provide an eloquent, grounded testament to Jennifer's enduring contributions through their work and those students that her mentees continue to mentor. This volume is not just a tribute to Jennifer, but a gift to the evaluation field to be reminded of the values and commitments that strengthen our practice.

—**Debra Rog**
Vice President for Social Policy and Economics Research, Westat

Disrupting Program Evaluation and Mixed Methods Research for a More Just Society

A volume in
Evaluation and Society
Stewart I. Donaldson and Katrina L. Bledsoe, *Series Editors*

Evaluation and Society

Stewart I. Donaldson and Katrina L. Bledsoe, *Series Editors*

Disrupting Program Evaluation and Mixed Methods Research for a More Just Society

The Contributions of Jennifer C. Greene

edited by

Jori N. Hall
University of Georgia

Ayesha Boyce
Arizona State University

Rodney Hopson
University of Illinois at Urbana–Champaign

≡**IAP**

INFORMATION AGE PUBLISHING, INC.
Charlotte, NC • www.infoagepub.com

Library of Congress Cataloging-in-Publication Data

A CIP record for this book is available from the Library of Congress
http://www.loc.gov

ISBN: 979-8-88730-104-4 (Paperback)
       979-8-88730-105-1 (Hardcover)
       979-8-88730-106-8 (E-Book)

Printed in the United States of America

CONTENTS

SECTION I

DEMOCRATIZING INQUIRY: FOSTERING INCLUSION, COLLABORATION, AND LEARNING

SECTION II

DIVERSIFYING SOCIAL SCIENCE INQUIRY NATIONALLY AND INTERNATIONALLY THROUGH PEDAGOGY AND MENTORSHIP

SECTION III

MEANINGFUL DIALOGUE: DISPOSITIONS, IMPOSITIONS, AND ASSUMPTIONS MATTER

SECTION IV

THE LAST WORD

PREFACE

At the time of this book's writing, many global crises are occurring, including but not limited to the COVID-19 pandemic, an economic recession, a resurgence of sociopolitical and racial unrest, and concerns surrounding climate change. These issues, while desperately pressing, are also opportunities to think more deliberately about the role of evaluators and researchers in disrupting the negative impacts of these issues. Disruption in this sense refers to how we, as evaluators and researchers, can position ourselves to think and respond in ways that can better describe and explain the nuance and complexity of social issues and other phenomena of interest. It speaks to how we can advance justice work that disrupts the harmful, destructive, or even fatal consequences that typically result from social problems—especially for those who are vulnerable, disadvantaged, or overlooked.

It is in this sense, we consider Jennifer C. Greene disruptive. That is, Greene has and continues to be committed to critically disrupting how phenomena and assumptions about the role of inquiry in society are traditionally considered. Specifically, her commitments are disruptive as they serve as a catalyst to *democratize* and enhance the quality of the inquiry process, promote a *diversity* of inquiry approaches and value the people served in the inquiry context, and personalize and reflect on the inquiry process through *dialogue*. Ultimately, Greene's value commitments advance analytical and empirical work to disrupt the negative consequences inequities can have on people. Given our current global context, we consider a book on Greene's commitment to values of democracy, diversity, and dialogue timely to think

Disrupting Program Evaluation and Mixed Methods Research for a More Just Society, pages xi–xii
Copyright © 2023 by Information Age Publishing
www.infoagepub.com
All rights of reproduction in any form reserved.

more intentionally, creatively, and boldly about how to disrupt the status quo and address real-world problems within the context of one's work.

CHAPTERS IN THIS BOOK

The book's chapter authors include program evaluators working in academic, educational, philanthropic, and government agencies; researchers located in nonprofit, corporate, and other firms specializing in interdisciplinary ideas; and university graduate students and instructors. These authors have been aware of the work of other contributors. And, in some cases, authors have been in direct communication with one another, generating rich interdisciplinary and intergenerational collaborations, resulting in the content of each section. While not all of the authors share the same perspective regarding Greene's contributions nor the characterization of her as disruptive, each author in this book has considered her impact on the fields of evaluation and mixed methods.

ACKNOWLEDGMENTS

We are thankful to all of the contributors to this book. In the midst of a pandemic and other challenges, you took time out of your schedules to write and share how Jennifer C. Greene impacted your personal and professional lives. We are grateful for each and every one of you.

We are indebted to the series editors who saw the grand vision of this book and supported the same by offering guidance throughout the writing and publishing process. Because of your support, this book will serve to increase the impact and significance of Greene's work.

We are grateful to each person who reads this book. We trust the book contents will be meaningful and thought-provoking in ways that inspire creativity and imagination in the context of evaluation and mixed methods.

INTRODUCTION

Values, Legacy, History, Practice, and Critique

The first two chapters in this introductory section provide an overview of Greene's value comments and describe a key institutional influence on Greene's work. In Chapter 1, Jori N. Hall and Janie Copple outline Greene's value commitments, providing a foundation for engaging subsequent chapters in this book. In Chapter 2, Rodney Hopson and Molly Galloway discuss the "Illinois School," the faculty and students who have contributed to evaluation scholarship for multiple decades at the University of Illinois in Urbana-Champaign. These authors highlight Greene's contribution to the Illinois School with special attention to her educative, deliberative, and social advocacy scholarship. The remaining chapters offer contrasting views on Greene as a disrupter in the fields of mixed methods and evaluation. In Chapter 3, Melvin M. Mark characterizes Greene as a disruptor and demonstrates how she shifted traditional mixed-methods conceptualizations and purposes for mixing qualitative and quantitative methods. In contrast, in Chapter 4, Thomas Schwandt, building on Greene's democratic evaluation scholarship, characterizes Greene as an interpretivist-humanist with a sensitivity toward stakeholder participation, equity, and social justice.

CHAPTER 1

THE VALUE COMMITMENTS OF JENNIFER C. GREENE

Jori N. Hall
University of Georgia

Janie Copple
Georgia State University

ABSTRACT

The purpose of this chapter is to contextualize the contributions of Dr. Jennifer C. Greene, whose legacy as a scholar-practitioner continues to develop and improve social science inquiry. To do so, we show how Greene distinctively disrupts traditional evaluation and mixed-methods inquiry. We frame our discussion using the value commitments that have guided her work: diversity, dialogue, and democracy. We discuss each value commitment, focusing on how the value commitment is characterized in the literature and how the value commitment is implemented in practice. This chapter importantly situates Greene's contributions to program evaluation and mixed methods, making visible how diversity, dialogue, and democracy importantly frame and provide a justification for her work. To conclude the chapter, we briefly introduce the other chapters in the book.

Disrupting Program Evaluation and Mixed Methods Research for a More Just Society, pages 3–18
Copyright © 2023 by Information Age Publishing
www.infoagepub.com
All rights of reproduction in any form reserved.

The purpose of this chapter is to provide an understanding of Greene's commitments to democracy, diversity, and dialogue. To begin, we discuss a foundational assumption that drives Greene's advancement of these commitments: *Inquiry is an inherently value-laden enterprise.* Then, we discuss each value commitment with attention to how it disrupts traditional ways of theorizing and practicing inquiry. Before doing so, we first share who we are and how we have come to know Greene and her work.

Briefly, the first author, Jori N. Hall, came to know Greene in graduate school at the University of Illinois in Urbana-Champaign (UIUC). In addition to enrolling in Greene's courses on evaluation theory and practice, Hall, along with others, worked with Greene to conceptualize and pilot the values-engaged, educative (VEE) evaluation approach (Greene et al., 2006). Since Hall graduated from UIUC, she has continued to write and work with Greene, contributing to the literature on the consequences of mixed-methods theory (Greene & Hall, 2010), values-engagement in program evaluation (Hall et al., 2012), alternative approaches to presenting evaluation findings (J. Johnson et al., 2013), and the diversity of mixed-methods inquiry approaches (Hall & Greene, 2019). Hall's long-standing professional and personal relationship with Greene has greatly influenced her use of theories and methods to disrupt privilege in evaluation (Hall, 2020), engage stakeholders' cultural values more responsively (Hall, 2020; Hall et al., 2020), and enhance the credibility of mixed-methods evaluation (Hall, 2013). As a professor in the qualitative research and evaluation methodologies (QREM) program at the University of Georgia (UGA), Hall teaches courses on qualitative research, evaluation theory, and mixed-methods research. In this capacity, she trains evaluators and supervises mixed-methods dissertation work, incorporating the VEE approach and other context-specific theories and methods.

The second author, Janie Copple, was introduced to Greene's work in Hall's evaluation theory course as a PhD student in the QREM program. With a background in middle grades education and research interests in puberty/sex education, Copple was initially inspired by Greene's VEE framework and excited to consider the possibilities of VEE approaches for future evaluations of K–12 sex education programs. Copple's research on adolescents coming of age engages issues to do with formal and informal puberty education and incorporates creative, multimodal approaches (e.g., narrative, poetic, arts-based) to qualitative data collection and representation. Greene's (1994) advocacy for qualitative inquiry in program evaluation, incorporation of qualitative methods in VEE (Greene et al., 2006), and alternative approaches for representing results in program evaluation (J. Johnson et al., 2013), inspires Copple to consider how she might incorporate multimodal qualitative data with VEE approaches in future program evaluations. Specifically, Copple is interested in the intersection of alternative, creative representations of data with Greene's (1994) pivotal question: "What values or whose?" (p. 540) in program evaluation. As Copple begins her career as assistant professor of qualitative research methods in the Department of Educational Policy Studies at Georgia State University, she hopes to spark discussions around such questions as she mentors future education researchers and evaluators. In writing this chapter, Copple and Hall collaboratively reflected on the major impact of Greene's disruptive thinking and empirical work on their inquiry practice, and the fields of program evaluation and mixed-methods research. This chapter presents a deeper exploration of how Greene has disrupted established norms through her value commitments based on those reflections.

ADVOCATING VALUES: DEMOCRACY, DIVERSITY, AND DIALOGUE

We begin our discussion on Greene's value commitments by briefly describing the debate in the evaluation community concerning whether evaluators should take on an advocacy role in evaluation. Starting with this debate is fitting because it exemplifies how Greene disrupted the traditional view of evaluation as value-neutral. Her thinking in this case disrupts the status quo by reframing the advocacy issue, suggesting that evaluation is inherently a value-laden enterprise.

The role of advocacy has been (Scriven, 1994; Stake, 1997) and remains (Boyce, 2019) controversial in the field of evaluation. Central to this controversy is the view that advocacy for or against a particular cause, policy, or program renders program evaluation biased and unscientific, thus violating traditional scientific ideals of program evaluation: objectivity and value neutrality. At the center of the controversy is the question: "Should evaluators take on an advocacy role in their evaluation practice?" Greene (1997), taking the controversy surrounding advocacy in program evaluation head-on, argues that advocacy in evaluation is unavoidable. This is because evaluators make value claims regarding what is important in the context of a particular evaluation. By doing so, evaluators advocate for particular questions, concerns, theories, and stakeholders to prioritize in their evaluation practice. In this way, evaluation practice inherently advocates specific values or principles. For instance, the empowerment evaluation—an approach that helps stakeholders conduct evaluation for program development purposes—advocates the principles of inclusion, organizational learning, and social justice (Fetterman & Wandersman, 2005).

Acknowledging both the tension in the evaluation community concerning whether evaluators should take on an advocacy stance and the values evaluators inevitably advocate through their evaluation practice, Greene (1997)

ultimately argues that evaluation is indeed advocacy. She posits, *"The important question then becomes not, should we or should we not advocate in our role as
evaluators, but rather what and whom should we advocate for?"* (p. 26, emphasis
added). Greene's reframing of the question concerning the role of advocacy
in evaluation supports her explicit and steadfast commitment to democracy,
diversity, and dialogue. These are the values she explicitly advocates. They are
made manifest intentionally and vigorously through Greene's scholarship,
as well as her service and relationships with graduate students and faculty
within and beyond the United States. As a result, Greene's values continue to
strengthen the quality and character of social science inquiry.

A VISIONARY VALUES FRAMEWORK

Greene's commitment to democracy, diversity, and dialogue reflect a visionary values framework. In what follows, we introduce each value commitment.
We view outlining Greene's value framework as important to demonstrate
how she disrupts traditional social science inquiry ideals (e.g., objectivity,
value neutrality) and enhances the quality of social science inquiry. Also,
Greene's values correspond to the three sections framing this book; therefore, we consider discussing each value commitment and the relationships
among them in this chapter as a useful entry point to engage the ways chapter contributors discuss the value commitments in subsequent chapters.

Relying on Greene's scholarship and some of the methodological literature
that influenced her work (i.e., Cronbach, 1983; Freire, 1970; House & Howe,
1999; MacDonald, 1978), we describe each value commitment and illustrate
how it is applied in practice. While Greene does not suggest a hierarchical
relationship between values, we argue she advocates for dialogue and diversity in service to democracy. With this in mind, we begin by discussing how
Greene's commitment to democracy disrupts traditional evaluation by legitimizing stakeholders' values, democratizing the inquiry process, and creating
opportunities for stakeholder decision-making in the evaluation process.

Disruptive as Democratizing the Inquiry Process

Greene's commitment to democracy builds on House and Howe's
(1999) deliberative democratic evaluation and other democratic-oriented
approaches to evaluation (MacDonald, 1978). A deliberative democratic
orientation requires including the interests of all relevant stakeholder
groups—especially those who lack power or who have been historically
marginalized or underrepresented. Extending the deliberative democratic evaluation approach, Greene (1997) advances the notion of *democratic*

pluralism. The notion of "democratic" in democratic pluralism acknowledges the importance of including all relevant stakeholders in dialogue in the context of an evaluation, with particular attention to those on the margins of society. The notion of "pluralism" suggests that there is more than one approach to truth, recognizing "the legitimacy of other value stances" (Greene, 1997, p. 26). Pluralism also requires power sharing in evaluation (Greene, 1997). That is, the evaluator has the responsibility but not the sole authority to make decisions about the program's quality. Democratic pluralism, then, requires opportunities for stakeholders to participate in decision-making processes about the program's merit or worth.

It should also be noted that democratic pluralism has implications for the role of the evaluator, positioning her as a "public scientist" (Greene, 1997, p. 29). Greene's use of the term "public" signals the evaluator's role is to use evaluation as a tool to promote democratic ideals of inclusive dialogue and participation for a more equitable society. Here, we note the relationship between the evaluator role/evaluation inquiry and society and the disruptive nature of this value commitment. Similar to other democratic scholars (MacDonald, 1978), Greene (1997) posits evaluation plays a major role in continuing public conversations in the context of an evaluation. Positioning evaluation this way disrupts the traditional role of the evaluator, moving her from an objective and distant scientist to a more engaged scientist, committed to promoting inclusive dialogue and representing multiple views. Thus, the public scientist is like a boundary spanner, linking evaluation inquiry to public engagement.

Democracy in Practice

Greene's commitment to democratic pluralism is reflected in her contributions to the field of program evaluation. For example, her belief in the legitimacy of value stances is embraced in two interrelated ways. First, it is recognized in her VEE approach to evaluation, which unapologetically prescribes democratic values such as inclusion (Greene et al., 2006). VEE, as will be explained in further detail in other book chapters (see, e.g., Boyce & Rivera, Chapter 9, this volume), originated in the context of science, technology, engineering, and mathematics (STEM) educational programs (Greene et al., 2006). This approach has moved across disciplines, encouraging its use in various contexts and the incorporation of different theories. For instance, VEE has been applied in the field of recreation and leisure studies to evaluate a community-based healthy aging program for older adults (Jaumot-Pascual, 2018) and combined with culturally responsive evaluation theory to evaluate educational reform efforts (Freeman & Hall, 2012; Hall et al., 2020). Second, democratic pluralism is advanced in

Greene's openness to alternative forms of data collection and reporting to ensure stakeholders' voices are included (J. Johnson et al., 2013). Alternative presentations (e.g., poetry, drama, music) legitimize diverse ways of presenting data, invite multiple interpretations, make data more accessible, provoke deep reflection, and reduce language barriers. Democratizing inquiry disrupts traditional notions of who participates and how stakeholders participate in knowledge production and decision-making, which has implications for what counts as knowledge, data, or evidence in evaluation and research. As a result, Greene's push to democratize inquiry expands the range of relevant standpoints and restructures power relationships, which increases the democratic potential of evaluation practice.

The Pitfalls and Promises of Deliberative Democratic Evaluation

We recognize Greene's commitment to democratic pluralism promoted in her democratic deliberative evaluation of a high school science education reform effort, focusing on the inclusion of non-honors students in honors science classes (Greene, 2000). Greene faced multiple challenges when enacting her democratic commitment and provided three reasons for this. The first involved the evaluation team's difficulty including community members (e.g., African American, Latinx, and low-income White parents) who were in favor of the reform initiative. The voices of these stakeholders were not well represented in the evaluation. Greene admits their lack of involvement was mainly due to recruitment fatigue and competing demands on the evaluation team's time. As a result, the voices of more powerful stakeholders who opposed the initiative were overly represented, thereby not fulfilling the "equity of voice" aspirations of her democratic-oriented evaluation (Greene, 1997, p. 19). Second, despite attempts to capture diverse viewpoints on the science classes and related values, opponents of the reform effort were preoccupied with the methods used for the evaluation. That is, the most vocal stakeholders demanded the evaluation rely on quantitative methods and outcomes (achievement test scores). Space for deliberation about the content of the science courses, the use of different methods, and the valuing of other outcomes was restricted, thereby reducing the evaluation's deliberative potential. The third challenge discussed concerned the authority of the evaluation. Greene's (2000) demands for pluralism were perceived by stakeholders as advocating for a "particular program position" (p. 24) rather than the inclusion of less powerful stakeholder voices, which, in turn, threatened the trustworthiness of the evaluation. Further, there was high administrative turnover at the high school, limiting support for the evaluation. Both Greene's perceived bias and lack of administrative support for the evaluation severely undercut the authority of the evaluation.

We point out how Greene (2000) discussed the challenges she faced when democratizing evaluation as it alerts evaluators to the realistic barriers and pitfalls that can occur in the context of democratic-oriented evaluation. Her concerns have been taken seriously by practicing evaluators and used to guide their evaluation practice. For example, using Greene's concerns to reflect on issues realized in their evaluation, Hreinsdottir and Davidsdottir (2012) implemented the deliberative democratic evaluation approach to evaluate Icelandic preschools. Similar to Greene, Hreinsdottir and Davidsdottir faced time constraints when employing their evaluation. Yet, they were able to include the most vulnerable stakeholders—preschool children—in their evaluation. While their evaluation was inclusive, Hreinsdottir and Davidsdottir note some participants were "anxious about discussing their views openly from fear of conflicts of interests" (p. 534) during their group discussions. Despite this, the information they gathered exposed "inequalities and injustices in the community and its institutions" (p. 534), which informed decisions about how the evaluation findings would be used. These evaluators came to realize the "preconditions for deliberative evaluation necessitate that the discussion groups be committed to the interests of the public, as without that commitment individual interests are likely to dominate" (p. 525). Ultimately, Greene's concerns informed Hreinsdottir and Davidsdottir's facilitation of vulnerable participants in the evaluation, identification of injustices, and collective decision-making processes. It also enabled these scholars to better understand the constraints and requirements necessary to actualize deliberative evaluation, namely stakeholders feeling secure enough to openly and honestly deliberate.

In short, Greene's commitment to democratize inquiry is disruptive in the sense it shifts thinking about evaluation from a value-neutral position to a value-laden enterprise. In doing so, it positions evaluation as a mechanism to intervene on issues of equity and power in society. Her democratic commitment, while ideal, fully acknowledges the practical and political issues associated with democratizing evaluation, particularly the inclusion of those most often excluded in evaluation practice. Greene's democratic goals enhance the quality of evaluation, providing diverse understandings, which assist in more accurate evaluative judgments. The importance of including different perspectives influences Greene's commitment to diversity, which values and encourages the use of different inquiry approaches, and the diversity of people served in the inquiry context.

Disruptive as Diversifying Difference

To characterize diversity, Greene draws on the work of Hazel Symonette (2004), who states, "Diversity includes those significant dimensions of human difference that have patterned implications for interpersonal relations and the nature of the interface with organizations, institutions, and other aspects of social structure" (pp. 2–3). Building on Symonette's (2004) definition of diversity, Greene acknowledges the multidimensional nature of human difference, recognizing the diversity of power and privilege, and the real-life consequences they have for how people are treated in particular contexts (Orellana & Bowman, 2003). Given the complexities of diversity, Greene et al. (2011) argue it is imperative to identify how diversity is constructed within a particular context in social science research. They further contend "working to understand diversity within the program context…is engaging equity" (p. 70). This argument is central to understanding how Greene disrupts traditional thinking about diversity. That is, Greene goes beyond thinking about diversity in terms of the demographics represented in a particular inquiry context. Rather, she links diversity to equity, suggesting that identifying important dimensions of difference is a prerequisite to participating in generative dialogue about inequities and gathering actionable information to address inequities (Greene et al., 2011). Further, Greene's notion of diversity not only meaningfully engages traditional categories of human difference (e.g., race, ethnicity, social class) but also considers diversity with respect to multiple paradigmatic perspectives, disciplinary viewpoints, and methodological approaches.

Diversity in Practice

Greene's commitment to understanding context-specific diversity is evidenced in the implementation of the VEE framework. Greene et al. (2011) further contend, "This evaluation approach aims to understand the ways in which diversity is constructed within the program in order for the dialogue on equity to be generative, contextually relevant, and meaningful" (Greene et al., 2011, p. 70). Here, we note how Greene promotes examinations of diversity in STEM educational programming as a means to discuss and address issues related to equity. In the VEE approach, equity concerns how program stakeholders are treated with special attention to program access, meaningful program engagement, and opportunities for achievement (Hall et al., 2012). Further, VEE demands that evaluators and stakeholders collaboratively conceptualize evaluation questions, collect information, and promote conversations that address how a program engages stakeholders and provides access to program content and activities, especially for groups who have been traditionally marginalized (Hall et al., 2012). Recently, STEM educational programming, research, and evaluation initiatives have given more attention to issues of diversity and equity (National Science Foundation, 2016); however, Greene's focus on diversity and equity in STEM evaluation was relatively pioneering at the time she and her colleagues developed and tested the VEE approach.

Engaging With Difference as a "Mixed-Method Way of Thinking"

As previously stated, Greene's (2007) notion of diversity not only includes demographic categories but also refers to her respect for multiple paradigmatic perspectives, disciplinary viewpoints, and methodological approaches that can be used to conceptualize, implement, and justify social science inquiry—or what she terms a "mixed methods way of thinking." According to Greene, a mixed methods way of thinking "rests on assumptions that there are multiple legitimate approaches to social inquiry and that any given approach to social inquiry is inevitably partial" (p. 20). In brief, a mixed methods way of thinking welcomes diversity of thought and different ways to make sense of the world. Below, we discuss two ways Greene's mixed methods way of thinking extends diversity.

First, Greene's mixed methods way of thinking contributes to the paradigmatic issue in the mixed-methods community about the legitimacy of mixing philosophical assumptions in the context of a single mixed-methods study. At the heart of this controversy is the argument that the assumptions of different paradigms are incommensurable. Greene rejects this argument and asserts the thoughtful integration of paradigmatic perspectives brings substantial potential to the sustained growth of mixed-methods inquiry in purposeful and innovative ways (Greene & Hall, 2010). Greene's way of thinking regarding the generative potential of mixing paradigms in a single study is most pronounced in her notion of the *dialectic stance.* More discussion on the dialectic stance is provided in our section on Greene's commitment to dialogue (e.g., "Disruptive as Dialogic Engagement").

Second, Greene's way of thinking contributes to methodological diversity, which is evidenced in her co-edited special issue on mixed methods in the *International Journal of Research and Method in Education* (Hall & Greene, 2019). Scholars featured in this special issue disrupt traditional "mixing" in mixed-methods inquiry by going beyond solely combining quantitative and qualitative methods. For example, in addition to mixing critical theory and quantitative criticism frameworks, Garnett et al. (2019) mix youth participatory action research with restorative practices to address inequities within a school district.

Central to methodological diversity is "mixing" with intention. This means whatever is being mixed—standpoints, methods, methodologies, paradigms—is done toward a purposeful objective or goal—not for the sake of mixing itself. Greene et al. (1989), identified five purposes for mixed-methods inquiry (e.g., complementarity, development, expansion, initiation, and triangulation) based on their analysis of 57 mixed-methods evaluation studies from the late 1970s and early 1980s, mostly from the United

States. Their empirical work, now over 30 years old, still contributes to the fields of mixed methods (Collins et al., 2006; Creamer, 2018) and evaluation (Christie & Fleischer, 2010), charting a path toward purposeful and meaningful social science inquiry.

Overall, Greene's commitment to diversity embraces the complexities and contextual situatedness of diversity and demands thoughtful engagement with human difference and the diversity inherent in social science inquiry itself. Her conceptualization of diversity importantly includes a mixed methods way of thinking that invites innovative and creative ways to theorize and investigate phenomena of interest. In these ways, Greene's approach to diversity is decidedly humanistic and holistic. She views engaging the difference diversity makes as necessary to democratize social science inquiry. Greene (2001) also acknowledges that engaging diversity requires dialogue, deep listening with respect for the uniqueness of the views offered, and the ability to resist the urge to privilege sameness or diminish divergence.

Disruptive as Dialogic Engagement

Greene's understanding of dialogue, to some extent, is informed by Freire (1970) and House and Howe (1999). Similar to these scholars, Greene (2001) posits dialogue requires authentic relationships and defines dialogue as "a value commitment to *engagement*" (p. 181, emphasis in original). For Greene (2001), dialogue also includes a commitment to equity. She makes clear that dialogue "involves moral relationships of trust, respect, caring, openness to difference, and political relationships of commitment to equity of power and voice" (p. 182). Ultimately, dialogic engagement is both a process and an outcome as it includes furthering authentic and inclusive conversations about difference and equity to produce defensible findings.

We observe dialogic engagement in Greene's dialectic stance (Greene & Hall, 2010), which

> actively welcomes more than one paradigmatic tradition and mental model along with more than one methodology and type of method, into the same inquiry space and engages them in respectful dialogue one with the other throughout the inquiry. (p. 124)

As mentioned earlier, this stance was put forward in response to a debate regarding paradigmatic commensurability in mixed-methods inquiry and disrupts the argument that different paradigms are incommensurable by assuming alternate paradigms can be in dialogue with one another. The dialectic stance importantly contributes to mixed-methods inquiry because it helps inquirers to (a) grapple with the complexity of phenomena, (b) consider what different paradigms make visible in the inquiry process, and (c) recognize and think with difference.

Dialectic Stance in Practice

As researchers grapple with how to incorporate multiple paradigms in a single inquiry project, they look to Greene's (2007) work as a guidepost for thinking with difference in mixed-methods research. For instance, in an editorial entitled "Living With Tensions: The Dialectic Approach," R. Burke Johnson (2008) outlined the ontological, epistemological, and methodological contentions surrounding a dialectic approach to mixed-methods research. Citing Greene's call for a dialectic stance to mixed-methods research, Johnson argued: "Qualitative, quantitative, and mixed research are three current webs of assumptions, beliefs, and practices that are dynamic and flexible across persons and groups that coalesce because they help researchers understand their worlds" (p. 206). More recently, R. B. Johnson (2017) has called for a dialectic pluralism metaparadigm in mixed-methods inquiry following Greene's assertion that "different people with different mental models contributing ideas to research, programs, projects and practice" (p. 159) is essential to furthering a democratic deliberative, justice-oriented process.

Building on Greene's work, other mixed-methods scholars have engaged conversations surrounding the tensions and possibilities of a dialectic stance. Cronenberg and Headley (2019), for instance, adopted an auto-ethnographic "dialectic dialogue" (p. 268) approach to discuss how dialectic stances shaped their dissertation research projects. As the authors dialogued about their journey toward a dialectic stance, they described initially feeling ontologically and epistemologically "boxed in" (p. 273) by singular methodological paradigms. They noted that dialectic stance allowed them to free themselves paradigmatically. They cautioned the dialectic stance does not imply an "anything goes" approach to dialectic, mixed-methods research. Rather, dialectic engagement requires careful, intentional attention to difference. Their application of the stance suggests the strength of the dialectic stance is an openness to inquiry that precludes prescriptive theoretical and methodological approaches.

The What and How of Dialogue

In addition to the use of the dialectic stance in mixed methods, Greene's commitment to dialogue is notable in her evaluation scholarship. In an article outlining VEE evaluation in STEM programs, Greene et al. (2006) pointed to the importance of dialogue in improving quality and access to

STEM education for underrepresented students. They drew upon Cronbach's (1983) "Ninety-five theses," recognizing the role of open and critical discussions toward an understanding of barriers to educational access and equity. Greene's VEE work in the context of STEM evaluation also provides some insights on how to engage dialogue in practice. However, it is important to note that Greene's work does not provide a "how-to" approach for facilitating critical dialogue, which could be considered a limitation, particularly for novice evaluators or seasoned evaluators new to VEE. However, refusing a prescriptive approach allows evaluators to remain nimble and responsive within program contexts. To that end, intentional and critical, values-engaged dialogue can look a variety of ways. How then are values brought to the table? Below, we provide a few examples.

In Greene et al.'s (2011) VEE guidebook, the authors note how evaluators might ask stakeholders to "articulate and critically reflect on their own experiences, perspectives and assumptions" (p. 81) regarding the education program and draw upon this feedback to cultivate "opportunities for dialogue and exchange" (p. 59). Additionally, guided by Carol Weiss's (1972) writing on program theory and evaluation, Greene et al. (2006) acknowledged how divergent program theories of change are fruitful places to engage dialogue among diverse stakeholders. Investigating and discussing stakeholders' program theories of change affords an opportunity to educate the evaluator as well as diverse stakeholders about multiple priorities and values surrounding a program.

In reflections on VEE in STEM program evaluation across three universities, Boyce (2017), following Greene et al.'s (2011) VEE guidebook, notes the importance of initiating "conversations about diversity and equity at the onset" of an evaluation (p. 37). Boyce (2017) also acknowledges the importance of inclusive dialogue among stakeholders in crafting evaluation questions, ensuring the perspectives of multiple stakeholders are addressed throughout the evaluation. Hall et al. (2012) underscore the need for values-engaged questions to facilitate dialogue among stakeholders who may hold differing opinions. For instance, asking stakeholders why a commitment to a particular value is favorable or using different data sets (narratives or interview excerpts from multiple stakeholder groups) to prompt conversations around diverse standpoints. Further, just as it is important to focus dialogue centered on equity and inclusion at the onset of an evaluation, it is equally important to return to these values when presenting evaluation results (Boyce, 2017). Moreover, Johnson et al. (2013) note how alternative presentations of evaluation results such as art, poetry, and performance in the presentation of evaluation results are useful to open discussion among diverse audiences and encourage critical reflection upon program goals, both essential for VEE evaluation.

Last, to facilitate inclusive dialogue, evaluators have played the role of a critical friend (Greene et al., 2011; Rallis & Rossman, 2000), one who views dialogue with others as a means to generate knowledge and demonstrates a willingness to "question the status quo and demand data to guide ethical decisions about change" (Rallis & Rossman, 2000, p. 84). Taking advantage of time for informal dialogue is essential in cultivating one's role as a critical friend. Prioritizing time requires "sustained engagement" and "respectful patience" (Boyce, 2017, p. 40). Evaluators must invest time to become educated on the dynamics of a program—learning how and when opportunities arise to guide decisions about change while maintaining respect for the diversity of opinions and experiences among stakeholders. In this way, VEE dialogue functions as both process and outcome, both iterative and responsive.

In sum, engaging in dialogue with those in the inquiry context is context-dependent. While there is no recipe for facilitating dialogic exchange, some options include using program theories, asking values-engaged questions, presenting evaluative findings using alternative approaches, and taking on the role of a critical friend. Greene's commitment to dialogue disrupts traditional purposes of evaluation (i.e., accountability) by using evaluation for a more educative function. That is, articulating stakeholder perspectives affords an opportunity to educate the evaluator as well as diverse stakeholders about multiple contexts, priorities, theories, values, and so on surrounding a program.

CONCLUSION

Although we addressed democracy, diversity, and dialogue in separate sections, these values are entwined and critical to respect diverse perspectives and chart a path toward more equitable, socially just inquiry outcomes (Greene, 2001). Greene's commitment to democracy, diversity, and dialogue reflects a visionary values framework that acknowledges the programs/research within which we work, and the values we hold are always, already contextually entangled. It is a framework that recognizes the multifaceted context surrounding difference and attends to the silences as well as the noise. Greene's values framework begins in service to democracy, realizing that social programs and social science inquiry are socially just when diverse experiences, backgrounds, and opinions are not only invited to the table but essential to the conversation. It is a framework that considers dialogue as something more than "just talk," but rather an intentional recognition of and reckoning with difference—among stakeholders, paradigms, and intellectual traditions. Thus, Greene's commitments bring critical conversations about diversity and equity into the public discourse. In doing so, she makes her vision clear not only for the type of inquiry we want to conduct but also for the type of society we want to create.

REFERENCES

Boyce, A. S. (2017). Lessons learned using a values-engaged approach to attend to culture diversity, and equity in a STEM program evaluation. *Evaluation and Program Planning, 64,* 33–43. http://doi.org/10.1016/j.evalprogplan.2017.05.018

Boyce, A. S. (2019). A re-imagining of evaluation as social justice: A discussion of the Education Justice Project. *Critical Education, 10*(1), 1–19. http://doi.org/10.14288/CE.V10I1.186323

Christie, C. A., & Fleischer, D. N. (2010). Insight into evaluation practice: A content analysis of designs and methods used in evaluation studies published in North America evaluation-focused journals. *American Journal of Evaluation, 31*(3), 326–346. https://doi.org/10.1177/1098214010369170

Collins, K. M. T., Onwuegbuzie, A. J., & Sutton, I. L. (2006). A model incorporating the rationale and purpose for conducting mixed methods research in special education and beyond. *Learning Disabilities: A Contemporary Journal, 4*(1), 67–100.

Creamer, E. G. (2018). *An introduction to fully integrated mixed methods research.* SAGE.

Cronbach, L. J., & Associates. (1983). Ninety-five theses for reforming program evaluation. *Evaluation Models: Evaluation in Education and Human Services, 6,* 405–412. https://doi.org/10.1007/978-94-009-6669-7_24

Cronenberg, S., & Headley, M. G. (2019) Dialectic dialogue: Reflections on adopting a dialectic stance. *International Journal of Research & Method in Education, 42*(3), 267–287. https://doi.org/10.1080/1743727X.2019.1590812

Fetterman, D. M., & Wandersman, A. (2005). *Empowerment evaluation principles in practice.* The Guilford Press.

Freeman, M., & Hall, J. N. (2012). The complexity of practice: Participant observation and values engagement in a responsive evaluation of a professional development school partnership. *American Journal of Evaluation, 33*(4), 483–495. https://doi.org/10.1177/1098214012443728

Freire, P. (1970). *Pedagogy of the oppressed.* Seabury Press.

Garnett, B. R., Smith, L. C., Kervick, C. T., Ballysingh, T. A., Moore, M., & Gonell, E. (2019). The emancipatory potential of transformative mixed methods designs: Informing youth participatory action research and restorative practices within a district-wide school transformation project. International Journal of Research & Method in Education, 42(3), 305–316. https://doi.org/10.1080/1743727X.2019.1598355

Greene, J. C. (1994). Qualitative program evaluation: Practice and promise. In N. K. Denzin & Y. S. Lincoln (Eds.), *Handbook of qualitative research* (pp. 530–544). SAGE Publications.

Greene, J. C. (1997). Evaluation as advocacy. *Evaluation Practice, 18*(1), 25–35. https://doi.org/10.1177/109821409701800103

Greene, J. C. (2000). Challenges in practicing deliberative democratic evaluation. *New Directions for Evaluation, 2000*(85), 13–26. https://doi.org/10.1002/ev.1158

Greene, J. C. (2001). Dialogue in evaluation: A relational perspective. *Evaluation, 7*(2), 181–187. https://doi.org/10.1177/135638900100700203

Greene, J. C. (2007). *Mixed methods in social inquiry.* Jossey-Bass.

Greene, J. C., Boyce, A., & Ahn, J. (2011). *A values-engaged educative approach for evaluating education programs: A guidebook for practice.* University of Illinois.

Greene, J. C., Caracelli, V. J., & Graham, W. F. (1989). Toward a conceptual framework for mixed method evaluation designs. *Educational Evaluation and Policy Analysis, 11*(3), 255–274. https://doi.org/10.3102/01623737011003255

Greene, J. C., DeStefano, L., Burgon, H., & Hall, J. N. (2006). An educative, values-engaged approach to evaluating STEM educational programs. *New Directions for Evaluation, 2006*(109), 53–71. http://doi.org/10.1002/ev.178

Greene, J. C., & Hall, J. N. (2010). Dialectics and pragmatism: Being of consequence. In by A. Tashakkori & C. Teddlie (Eds.), *SAGE handbook of mixed methods in social and behavioral research,* (2nd ed., pp. 119–143). SAGE.

Hall, J. N. (2013). Pragmatism, evidence, and mixed methods evaluation. *New Directions for Evaluation, 2013*(138), 15–26. https://doi.org/10.1002/ev.20054

Hall, J. N. (2020). *Culturally responsive approaches for qualitative inquiry and program evaluation.* Myers Education Press.

Hall, J., Ahn, J., & Greene, J. C. (2012). Values engagement in evaluation: Ideas, illustrations, and implications. *American Journal of Evaluation, 33*(2), 195–207. https://doi.org/10.1177/1098214011422592

Hall, J. N., Freeman, M., & Colomer, S. E. (2020). Being culturally responsive in a formative evaluation of a professional development school: Successes and missed opportunities of an educative values-engaged evaluation. *American Journal of Evaluation, 41*(3), 384–403. https://doi.org/10.1177/1098214019885632

Hall, J. N., & Greene, J. C. (2019). A kaleidoscope of perspectives on the potential, contributions, and grand vision of a mixed methods approach to educational inquiry. *International Journal of Research & Method in Education, 42*(3), 223–224. https://doi.org/10.1080/1743727X.2019.1602984

House, E. R., & Howe, K. R. (1999). *Values in evaluation and social research.* SAGE Publications.

Hreinsdottir, A. M., & Davidsdottir, S. (2012). Deliberative democratic evaluation in preschools. *Scandinavian Journal of Educational Research, 56*(5), 519–537. http://doi.org/10.1080/00313831.2011.599426

Jaumot-Pascual, N. (2018). *Conducting a culturally responsive evaluation: Values, engagement, self-reflexivity, and photo elicitation* [Unpublished doctoral dissertation]. The University of Georgia. http://getd.libs.uga.edu/pdfs/jaumot-pascual_nuria_201808_phd.pdf

Johnson, J., Hall, J. N., Greene, J. C., & Ahn, J., (2013). Exploring alternative approaches for presenting evaluation results. *American Journal of Evaluation 34*(4), 486–503. https://doi.org/10.1177/1098214013492995

Johnson, R. B. (2008). Living with tensions: The dialectic approach. *Journal of Mixed Methods Research, 2*(3), 203–207. https://doi.org/10.1177/1558689808318043

Johnson, R. B. (2017). Dialectical pluralism: A metaparadigm whose time has come. *Journal of Mixed Methods Research, 11*(2), 156–173. https://doi.org/10.1177/1558689815607692

MacDonald, B. (1978, October 17). *Evaluation and democracy* [Public address]. University of Alberta Faculty of Education, Edmonton, Canada.

National Science Foundation. (2016). *Broadening participation in STEM.* Arlington, VA. Retrieved from https://nsf.gov/od/broadeningparticipation/bp.jsp

Orellana, M. F., & Bowman, P. (2003). Cultural diversity research on learning and development: Conceptual, methodological, and strategic considerations. *Educational Researcher, 32*(5), 26–32.

Rallis, S., & Rossman, G. (2000). Dialogue for learning: Evaluator as critical friend. *New Directions for Evaluation, 2000*(86), 81–92. https://doi.org/10.1002/ev.1174

Scriven, M. (1994). Evaluation as a discipline. *Studies in Educational Evaluation, 20*(1), 147–166. https://doi.org/10.1016/S0191-491X(00)80010-3

Stake, R. E. (1997). Advocacy in evaluation: A necessary evil? In E. Chelimsky & W. R. Shadish (Eds.), *Evaluation* for the *21st century: A handbook* (pp. 470–476). SAGE Publications.

Symonette, H. (2004). Walking pathways toward becoming a culturally competent evaluator: Boundaries, borderlands, and border crossings. *New Directions for Evaluation, 2004*(102), 95–109. https://doi.org/10.1002/ev.118

Weiss, C. H. (1972). *Evaluation research: Methods of assessing program effectiveness.* Prentice Hall.

CHAPTER 2

SITUATING JENNIFER C. GREENE'S EVALUATION LEGACY IN THE ILLINOIS SCHOOL OF EVALUATION

Rodney Hopson
University of Illinois–Urbana Champaign

Molly Galloway
University of Illinois–Urbana Champaign

ABSTRACT

This chapter explores Jennifer Greene's ongoing legacy and situates it within what we refer to as the Illinois School of Evaluation, the collection of the University of Illinois–Urbana Champaign (UIUC) faculty and students who have contributed to evaluation thinking and scholarship for over 5 decades. In the first section of the paper, we imagine and make connections between Greene and the Illinois School of Evaluation, one that would place Greene on its Mount Rushmore of evaluation pantheons that have made indelible

Disrupting Program Evaluation and Mixed Methods Research for a More Just Society, pages 19–33
Copyright © 2023 by Information Age Publishing
www.infoagepub.com
All rights of reproduction in any form reserved.

contributions to the evaluation field. Next, we reflect on our relationship with Greene and her influences on our evaluation practice as a cherished professor, mentor, and friend.

Melvin Hall visited the University of Illinois in October 2019. Hall is a Northern Arizona University emeritus professor and alumnus of our educational psychology department, which housed the same scholars and thinkers in the field of evaluation that we can refer to as the *Illinois School*—the collection of University of Illinois–Urbana Champaign (UIUC) faculty and students who have contributed to evaluation thinking and scholarship for over 5 decades. Hall's visit to the college was as much of a reunion as it was a homecoming and reminded us of Jennifer's many contributions and legacies in the field. In his informal conversation with current UIUC evaluation colleagues and students, Hall reminded us of the particular tradition of education he received through support from noted scholars at the Center for Instructional Research and Curriculum Evaluation (CIRCE), notably through Terry Denny, Tom Hastings, Bob Stake, and countless other graduate students, spouses, researchers, and evaluators in the early 1970s. Figure 2.1 shows a picture of some of the CIRCE family in 1981.

Figure 2.1 CIRCE, 1981. *Back row:* Eileen Broderick, Deb Rugg, Jim Raths, Deb Trumbull, Tom Hastings, Jim Wardrop, Janice Hand, Bernie Stake, Colleen Frost, Jim Pearsol. *Front row:* Marli Andre, Claire Brown, Bob Stake, Sandra Mathison, Joyce Jones, Gordon Hoke, Hallie Preskill

Figure 2.2   CREA picture of CREA Affiliates at 2016 CREA Conference in Chicago, IL. Left to Right: *Back row:* Kevin Favor, Karen Kirkhart, Rodney Hopson, Sharon Nelson Barber, Melvin Hall, Jennifer Greene, Stafford Hood, Fiona Cram, Henry Frierson. *Front row:* Katrina Bledsoe, Joan LaFrance, Carolyn Turner, Toks Fashola, Monica Mitchell, Kathy Tibbetts, Nicky Bowman

Hall was taught to be purposeful, value the collective struggle of ideas that lie in and out of the classroom, look beyond the implied details through methodological rigor, and never be complicit as an evaluator. Hall's reflection on the contributions of those we consider to be part of the Illinois School laid the groundwork for recent and contemporary scholars and thinkers reflected in the legacies of others such as Jennifer Greene, Tom Schwandt, Lizanne DeStefano, Katherine Ryan, Stafford and Denice Hood, and the development of the Center for Culturally Responsive Evaluation and Assessment (CREA). Hall reminded us that the Illinois School tradition is one that pushes the boundaries of evaluation toward a more equitable society and that its legacy is fueled by meaningful and purposeful scholarship. Figure 2.2 shows a picture of some of the CREA Affiliate Researchers at a CREA conference in 2016.

This book—and this chapter in particular—explores Greene's ongoing legacy through her former students, colleagues, and peers. This chapter begins by imagining and making connections between Greene and the Illinois School of Evaluation, one that would place Greene on its Mount Rushmore of evaluation pantheons that have made indelible contributions to the evaluation field. We close the chapter by situating our academic positionality and relationship to Greene, and because of her scholarship, tutelage, and friendship, consider ways to think about how our evaluation practices are becoming more educative, more democratic, more socially just, and more critical in how we apply our values-engaged, purposefully culturally responsive, equitable, and transformative realities and possibilities.

JENNIFER'S ONGOING LEGACY IN EVALUATION: CONTRIBUTIONS OF THE ILLINOIS SCHOOL

Greene's voluminous scholarship represents an important contribution to the legacy of the evaluation field and to the particular group of professors, students, and staff, housed primarily in the College of Education within its departments and affiliated centers, especially their ideas nourished and promoted at the UIUC over a 50- to 60-year period. The Illinois School, arguably the longest tradition of evaluation scholarship in the U.S. higher education, is distinct from evaluation programs, units, or divisions in other predominately White institutions of higher learning, past or present.

As a graduate student in the late 1980s and early mid-1990s at the Curry School at the University of Virginia, Hopson encountered a landscape of evaluation programs at universities different from the landscape of programs Galloway faces as a soon-to-be-minted PhD. What were well-established programs at Cornell University, University of Virginia, and Ohio State University have now folded or morphed. Programs housed at the University of Connecticut, Western Michigan University, Michigan State, University of Minnesota, University of North Carolina–Greensboro, and Claremont Graduate University emerged in the last decade to new ideas and roles of the evaluation and evaluators in our society.

Evaluation programs and units face the same sensitive and market-driven realities as any discipline or interdisciplinary study, reflected by the evolving and dynamic nature of governmental, philanthropic, and academic support for the field, the relevance and depth of the field and its professional associations to the ongoing study of education and social problems, the number and interest of trained scholars and practitioners able to assume positions in multiple sectors, as well as other factors that contribute to the production of academic faculty and emerging scholars and practitioners in the field. While the list and analysis of evaluation programs documented by John Lavelle (2018, 2020) provides an illustrative and contemporary perspective of program and course offerings, their departmental and college locations, and their credit-bearing and required offerings, it requires reading more than university catalogs and other publicly available sources[1] to gather a sense of the character, purpose, or traditions of schools of evaluation. To understand and distinguish schools and thoughts of evaluations includes understanding the faculty scholarship, their influence in the field, and other archival, written, or oral accounts of these same stewards of the discipline.

The Illinois School has cemented its place in the fundamentals of the discipline—its theory, method, practice, and profession (Smith & Brandon, 2007). In fact, on the current edition of the evaluation theory tree (Alkin & Christie, 2013), the Illinois School is represented on the use, methods, and valuing of the names on the roots and branches of the evaluation theory tree. For years, the Illinois School benefitted from the direct influences and protégé of educational and evaluation pioneers such as Ralph Tyler and John Dewey, two prominent educationists who inspired the birth and development of ideas that shaped the graduate schooling and academic evaluative learning at UIUC (Cronbach, 1998; O'Shea, 1985). To them, the Illinois School represented an unparalleled pursuit of democracy and democratic principles that might contribute to foundational knowledge and innovative techniques or procedures to solve the country's and world's most complex and perplexing social and educational problems.

If there were an occasion that would provide a foray into the Illinois School (and distinguish Greene's own scholarship), it would be at the symposium to honor Bob Stake in May 1998. Attending as a senior member of the Cornell faculty among the more than 250 scholars, students, family, and friends, Jennifer presented a paper—"Balancing Philosophy and Practicality in Qualitative Evaluation"—a not-so-apt title to more central issues in the field of evaluation. Arriving in May of 1998 to the Stake Symposium, by Fall of 1999, she returned to Urbana Champaign to assume the role as professor in the Department of Educational Psychology, joining Stake, Katherine Ryan, Lizanne Destafano, and Tom Schwandt (who would join the faculty with her the same semester from Indiana University–Bloomington). Her comments at Stake's festschrift lays out her own contributions to the field and the Illinois School; mostly the paper serves as harbinger to her role as disrupter, as catalyst, and as saboteur, building off the influences of her senior colleague on that spring occasion. Greene (1998) notes how Stake (and especially Cronbach) influenced her (and other fledgling evaluations in the 1970s) to come to a better understanding of the following three things:

1. "to make sense out of the mismatch between what I knew how to do—experimental designs and statistical analyses—and what was likely to be meaningful to those in the sites in which I was working;
2. "to begin to learn more about alternative ways to do evaluation and other forms of other applied social inquiry . . . about our social world and how we can know it; and
3. "to start to develop our self-consciousness about how our methods make statements about values, about value choices available to evaluators, and about the challenges in honoring multiple value stances and perspectives in any given evaluation." (pp. 35–36)

The paper in honor of Stake provides a version of what Greene would bring to the Illinois School, including adding to the core challenges of the field. She situates two challenges, one from the center and one from the edge, that would amplify her disruptive stance and align with the conceptual and methodological foundational underpinnings of the Illinois School.

Challenges from the center, "selected statements from the center evaluation community—on what we as evaluators should be doing these days" (Greene, 1998, p. 36) exemplify agencies, associations, and the technical aspects of government, for instance. More than a group of data technicians, analysts, or scientists, the contributions of Greene and others at the Illinois School call on evaluators to employ meaningfulness, deliberation, and purpose in pursuit of credible methods and measures in diverse settings. Moreover, their contributions call on stakeholders to recognize the role of evaluation for social betterment and social advocacy as an ethical, social, and political act (Greene, 2001a, 2009, 2010, 2012, 2014a, 2014b, 2016; Greene et al., 2007; Hall et al., 2012).

Challenges from the edge, "challenges to the very nature of our qualitative data, challenges to the meaning of our interpreted meanings, and challenges to the political location of our work" (Greene, 1998, p. 39) represent giving more honor to how meanings are constructed and by whom. At the core of these ideas influenced and incorporated in an educative, engaged tradition of evaluation, she extends our way of thinking about diversity and difference as core commitments to our work with and in service to social programs and our larger world (Greene, 1999, 2005a, 2014a). Her mixed-methods work is foundational and relevant to both the theory and practice of evaluation and beyond (Greene, 2001b, 2002a, 2002b, 2005a, 2005b, 2005c, 2005d, 2006, 2009b, 2015; Greene & Caracelli, 1997a, 1997b; Greene et al., 1989; Greene et al., 2011; Greene et al., 2010), more than mixing methods but the mixing of meanings with those most engaged and attentive to "cultural, racial, ethnic, linguistic, geographic, gendered political (and other dimensions) of a program context" (Greene, 2005a, p. 13).

GREENE'S INFLUENCE ON MOLLY'S EVALUATIVE THINKING AND PRACTICES OF HEALING, RECIPROCITY, POWER, AND SOCIAL JUSTICE

As a graduate student at the University of Illinois, I learned lessons from Dr. Greene that have helped shape my research agenda and motivations in profound ways. Since I was a child, I have had a burning desire to help people. I hated all the pain I saw in the world, and I wanted to be a part of the healing process. This desire took me down many different paths, trying to find the best way to aid in the healing of my community. During my undergraduate education, I started to view academia as the ideal career for aiding in the fight for social justice on both policy and individual levels. I had this glorified view of what it meant to be in academia. I believed it provided a space to conduct research that would help develop policies for the betterment of my community while also aiding in the mentorship of students with similar goals. However, after entering graduate school, I quickly realized that these values were not inherent to the work of academics. This realization reshaped my views of research and academy.

I wanted to conduct research for a purpose. I had no desire to sit in an "ivory tower" discussing social problems with little intention of addressing them. Thus, I not only wanted to gain more insight into the social problems that plagued the urban and rural communities like the ones I came from, such as food deserts, poverty, and mass incarceration, I also wanted to aid in efforts to eradicate them. As a nascent graduate student, I had only a vague understanding of evaluation. Evaluation was that thing my mother, a high school science teacher, was always stressing about at work. It wasn't until I had a conversation with a close friend that I realized evaluation was much more than a way to assess teacher performance. It was a way to produce research with the intention of the research being useful for the people who had a stake in the programs or policies being evaluated. Although I no longer had the hope that I once felt for academia and research being a tool to aid in social justice, I decided to give it one more try.

Searching for answers to my life goals and values, I enrolled in my first evaluation course, Evaluation Methods, in the Department of Educational Psychology. Little did I know that the course was taught by a professor that I would soon come to admire, Dr. Jennifer Greene. In that course, Dr. Greene taught us the basics of evaluation by providing us with an educational space to learn through practice. Having the opportunity to work with program stakeholders at every stage of an evaluation, along with the course readings, shaped my understanding of evaluation as a tool that can and should be used for the public good (Greene, 2010; Tarsilla, 2010). For example, I learned the importance of making reports accessible to different stakeholders by negotiating who has access to the reports and presenting the results in a manner that is responsive to the program context and various stakeholders (Johnson et al., 2013).

Over the course of the semester, I realized that Dr. Greene not only published on her commitments to envisioning a democratic society (Greene, 1997), to the values of equity and inclusion (Greene, 2001a, 2012), and on the importance of educating stakeholders, but about her values and evaluation in general (Greene et al., 2006). These commitments were also apparent in her practice and teaching. For example, one aspect of the course that has stuck with me over the years is the method she used to find evaluation projects for the methods course. Dr. Greene solicited local public schools for possible evaluation problems and projects with a two-fold purpose: (a) to provide the students in the course a space to learn evaluation through practice and (b) to provide the schools an opportunity to have a program evaluated for free with academically trained doctoral students (with lived experiences in education and social sciences), a faculty advisor, and an instructor. Dr. Greene stressed the importance of leaving a space better than how we found it, and she would remind us that it is important to remember that the communities where these programs exist will continue after we leave. By educating stakeholders and her purposeful acknowledgment and attempt to restructure power dynamics, she instilled in us a stance against oppressive practices and placed herself on the side of those traditionally ignored.

The lessons that I learned from Dr. Greene's teachings and scholarship, along with other prominent UIUC scholars and evaluators, such as Rodney Hopson, Stafford Hood, and Denice Hood, reignited my faith in the ability of research to serve as a tool for social change and helped to shape my research in profound ways. For example, my early evaluation experiences in Dr. Greene's courses provided me with an evaluation foundation that centers around my values and desire to work alongside grassroots activists, with the intention of creating a more equitable society. These experiences provided me with an introduction to evaluation practice that I now realize was based on reciprocity and was outside of the structure and power dynamics that tend to accompany evaluations that are attached to grants and funders. The relationship between our evaluation methods course and the schools was a reciprocal one, detached from grants and funders, in that they provided us the space to practice evaluation by giving us access, and we used evaluation to provide them feedback with the intention of being useful for them.

Over the years, I have continued to ponder ways to conduct evaluations that are not bonded by grants and funders but instead are built on a reciprocal relationship, like the one that brought me to the practice of evaluation, as I consider the potential benefits and barriers to doing this work. For instance, my current PhD research is a qualitative case study of a youth boxing program where I use evaluation as a form of reciprocity for the program and its participants, allowing me access to the space, the participants, their lives, and their stories.

GREENE'S INFLUENCE ON RODNEY'S UNDERSTANDING OF CULTURALLY RESPONSIVE EVALUATOR AS STEWARD OF PUBLIC GOOD AND DEMOCRATIZING SOCIAL JUSTICE PRACTICE

Although I have not benefited from being in Jennifer's classroom as a student, in other ways I have benefitted from the remarkable and indelible footprint that she has laid for so many of us in evaluation. I cannot exactly recall my first time meeting her (and with a failing memory, I am not sure it matters), but she most recently reminded me (again!) that we met in Puerto Rico.[2] Trust that while she recalls meeting me, I know for a fact that she was one of the evaluation theorists and scholars in the field that I had read about before she "met" me. Fast forward 20-plus years, Jennifer is very much a part of my professional and personal life. Like an academic auntie, she has played pivotal roles in guidance, support, and admonitions at key stages of my academic career. She has never been an auntie who wants a lot of credit, she is happy to be on the sideline, but she very much has left an indelible mark on the development of my career.

In her own tribute to her parents (Greene, 2018), I better understand the qualities she possesses and why I was drawn to her early in my career. She would write in the opening of a compendium of 29 evaluation lives (Williams, 2018),

> I think trying to make a difference came from my parents (Jim and Whit), my family upbringing—from both of my parents in different ways. They did not tell me this; I just learned it from them, from who they are, what they did, and how they acted. (Greene, 2018, p. 75)

She describes a service- and civic-oriented, giving, and generous father and mother who valued the ideals and practice of citizenship engagement and the importance of giving back and contributing to the public good, equality, and equity in our world.

After meeting and getting mentored through one of my first evaluation contributions, I looked for more opportunities to work with her. Shortly after Puerto Rico, I was invited along with other more senior colleagues by the Lucille and David Packard Foundation to a convening to facilitate ideas and practices between philanthropy and evaluation. Out of the convening, a book was published (Braverman et al., 2004), and more importantly, I had the good fortune to work with Jennifer and Ricardo Millett on a chapter. I recall being in awe of these two senior colleagues in the field and extending contributions to using evaluation as a democratizing practice (Greene et al., 2004).

That was a watershed moment for me, largely due to how much fun it was collaborating with Jennifer and Ricardo but also due to the safe space they shared in presenting ideas that resonate with me and my own upbringing in the field. In that initial co-publication, Jennifer and Ricardo helped to give wings to ideas I too valued: "So, the critical questions became, not only which evaluation approaches and methods should be used in a given context, but also which political positions and values should be advanced in the social practice of evaluation" (Greene et al., 2004, p. xx).

My practice of evaluation has and continues to recognize the (social) justice turn in evaluation. I credit Jennifer as foundational to my thinking about and envisioning evaluation as educative, democratizing, and as reinventing an evaluation that recognizes our relational responsibilities as evaluators (Hopson, 2005; Hopson & Cram, 2018). Like those she emulated (such as Griffith & Montrosse-Morehead, 2014; House, 1980; House & Howe, 1999) for their historical and contemporary contributions, Jennifer's work has provided a platform to consider a liberating role of evaluation, one that either maintains and reinforces a status quo or challenges or disrupts the existing social order (Greene et al., 2004).

My ongoing work in communities like Pittsburgh, PA, includes teams of researchers and evaluators who have built community and university relationships in ways to support the advancement of vulnerable communities through re-envisioning transformational leadership and social change (Hopson et al., 2016; Miller et al., 2011). Other ongoing work in Champaign, IL (Cooper, 2021), reflects Deweyan (1927), Freirean (1970, 1972), and Greenian educative visions of community–university partnerships.

Through the LIFT (Leading Individuals and Families to Transformation) Champaign project (Champaign School District, 2021), we evaluators in the CREA and in the EvaLab in the Department of Educational Psychology of the College of Education were invited by our chancellor's office to enter a scope of work with the City of Champaign and the Unit 4 Champaign School District earlier in 2021. Informed by LIFT's goal "to improve the lives of African American youth and their families by providing intentional wraparound support services and the necessary resources to build on their strengths and to help them achieve personal, academic, career, and interpersonal goals" (Champaign School District, 2021, n.p.), we are designing a series of phase-oriented evaluation activities (e.g., building a program logic model and a community asset map) and other measures intended to help determine the quality and effectiveness of the LIFT program guided by a culturally responsive evaluation framework (Hood et al., 2015). From the start, an educative and facilitative process with key stakeholders from the Champaign city and school district has led to the initial 9 months of relationship and trust building (a) in the successful passage of local school district and city council approvals, (b) in developing a memorandum of agreements and reviewing of evaluation and programs for the larger purpose of having serious engagement with participants and key stakeholders, and (c) in recognizing the context-sensitive nature of the work. As stewards of the public good, we recognize how important it has been to "visit" and "listen well"[3] with our colleagues in the administrative team made up of key directorates within the city/school collaboration and with the core project staff made up of champions, a family liaison, and other staff who work directly in the planning and implementing of the anticipated 70 LIFT Champaign families.

I thank Jennifer for helping me to ask critical questions about the role, place, and position that we evaluators advance in our practice and for helping me use my culturally responsive practices as a steward for public good and discourse whereby we might serve communities less privileged and underserved to purposefully and intentionally uplift their important voices of promise, hope, and equity.

CONCLUDING COMMENTS AND COMMITMENTS

We begin the concluding comments with Jennifer's words in her tribute to her parents, "In gratitude to Jim and Whit":

> I think the integration and the connection of those political beliefs and values with my evaluation work have developed over the years, because for quite some time, I did not see evaluation as a very politicized activity. But now I know evaluation is not this neutral process that just collects data neutrally and just reports objectively. It advances some values and not others. And once I realized that, then the question became, what values to advance? And that goes back to my parents in the 1960s, making sure that those who are least well served and those who are most disadvantaged in a particular context are those who are part of the data, part of the voices, part of what constitutes evidence in a particular context. That remains a commitment. (Greene, 2018, p. xx)

Part of the data. Part of the voices. Part of what constitutes evidence in a particular context. Within the sociopolitical milieu of our evaluation practice exist increasing tensions around matters of inequality, inequities, and disparities in education, social sectors, and services the world over. We, the authors of this chapter, recognize and pass Jennifer Greene's baton to the courageous and inspired evaluators to reflect on Greene's living scholarly legacy and contribute meaningfully to redefining traditions, theories, and practices of our discipline and profession.

ACKNOWLEDGMENTS

Thanks to those colleagues who offered suggestions to earlier versions of previous drafts by: Melvin Hall, Jori Hall, Ayesha Boyce, Rebecca Teasdale, Aneta Cram, and Amy Welde.

ENDNOTES

1. Interestingly, by virtue of the same changing dynamics noted earlier regarding the ebb and flow of evaluator education programs at universities, the UIUC campus was absent from the Lavelle (2018) publication.

2. The irony is that the first time that I visited Puerto Rico was for an applied anthropology conference, one that helped me shape the *Language Matters* (Hopson, 2000) issue in which she served as journal co-editor with Gary Henry. So, if Jennifer said she saw me there as she describes "a guy with more hair on his head" and a few other notable observations, I will accept it.
3. With references to 20th century Jewish philosopher Hannah Arendt (1958), Greene (2005a) refers to "visiting" and "listening well" as a "first set of ideas probes what it means to engage with difference in evaluation on the ground" (p. 15), intended to further aspects of respect, good judgment, and "learning about their lived experiences from your own eyes but within their stories in their spaces and places" (p. 16).

REFERENCES

Alkin, M., & Christie, C. (2013). An evaluation theory tree. In M.C. Alkin (Ed.), *Evaluation roots: A wider perspective of theorists' views and influences.* (2nd ed., pp. 11–57). SAGE. https://dx.doi.org/10.4135/9781412984157.n2

Arendt, H. (1958). *The human condition.* University of Chicago Press.

Braverman, M. Constantine, N., & J. K. Slater (Eds.) *Foundations and evaluation: Contexts and practices for effective philanthropy.* Jossey-Bass.

Champaign School District (2021). LIFT Champaign [Website]. https://www.champaignschools.org/families_and_students/lift

Cooper, R. (2021, November 4). *School district-wide program to LIFT up underserved families officially launches.* WCIA. https://www.wcia.com/news/local-news/school-district-wide-program-to-lift-up-underserved-families-officially-launches/

Cronbach, L. (1998). Saturday's opening session introductory remarks. In R. Davis (Ed.), *Proceedings of the Stake symposium on educational evaluation* (pp. 1–3). University of Illinois.

Dewey, J. (1927). *The public and its problems* (1st ed.). Swallow Press. https://doi.org/10.5840/monist193040413

Freire, P. (1970). *Pedagogy of the oppressed.* Herder and Herder.

Freire, P. (1972). Education: Domestication or liberation? *Prospects, 2*(2), 173–181. https://doi.org/10.1007/BF02195789

Greene, J. C. (1997). Evaluation as advocacy. *Evaluation Practice, 18*(1), 25–35. https://doi.org/10.1016/S0886-1633(97)90005-2

Greene, J. C. (1998). Balancing philosophy and practicality in qualitative evaluation. In R. Davis (Ed.), *Proceedings of the Stake symposium on educational evaluation* (pp. 35–49). University of Illinois.

Greene, J. C. (1999). The inequality of performance measurements. *Evaluation, 5*(2), 160–172. https://doi.org/10.1177/13563899922208904

Greene, J. C. (2001a). Dialogue in evaluation: A relational perspective. *Evaluation, 7*(2), 181–187. https://doi.org/10.1177/135638900100700203

Greene, J. C. (2001b). Mixing social inquiry methodologies. In V. Richardson (Ed.), *Handbook of research on teaching* (4th ed., pp. 251–258). American Educational Research Association.

Greene, J. C. (2002a). Is mixed methods social inquiry a distinctive methodology? *Journal of Mixed Methods Research, 2*(1), 121–140. https://doi.org/10 .1177/1558689807309969

Greene, J. C. (2002b). Mixed-method evaluation: A way of democratically engaging with difference. *Evaluation Journal of Australasia, 2*(2), 23–29. https://doi .org/10.1177/1035719X0200200207

Greene, J. C. (2005a). Evaluators as stewards of the public good. In S. Hood, H. Frierson, & R. Hopson (Eds.), *The role of culture and cultural context: A mandate for inclusion, truth, and understanding* (pp. 7–20). Information Age Publishing.

Greene, J. C. (2005b). Section II commentary: "The eyes of the beholders." *Early Education & Development, 16*(4), 489–492. https://doi.org/10.1207/s1556 6935eed1604_8

Greene, J. C. (2005c). Synthesis: A reprise on mixing methods. In T. S. Weisner (Ed.), *Discovering successful pathways in children's development: Mixed methods in the study of childhood and family life* (pp. 405–419). The University of Chicago Press.

Greene, J. (2005d). The generative potential of mixed methods inquiry. *International Journal of Research & Method in Education, 28*(2), 207–211. https://doi .org/10.1080/01406720500256293

Greene, J. (2006). Toward a methodology of mixed methods social inquiry. *Research in the Schools, 13*(1), 93–99.

Greene, J. (2009a). Evidence as "proof" and evidence as "inkling." In S. Donaldson, C. Christie, & M. Mark (Eds.), *What counts as credible evidence in applied research and evaluation practice?* (pp. 153–167). SAGE.

Greene, J. (2009b). Meaningfully engaging with difference through mixed methods educational evaluation. In K. Ryan & J. B. Cousins (Eds.), *The Sage international handbook of educational evaluation* (pp. 323–340). SAGE.

Greene, J. C. (2010). Serving the public good. *Evaluation and Program Planning, 33*(2), 197–200. https://doi.org/10.1016/j.evalprogplan.2009.07.013

Greene, J. C. (2012). Values-engaged evaluation. In M. Segone (Ed.), *Evaluation for equitable development results* (pp. 192–206). Retrieved from http://www.clear-la.cide.edu/sites/default/files/Evaluation_for_equitable%20results_web.pdf

Greene, J. (2014a). Culture and evaluation: From a transcultural belvedere. In S. Hood, R. Hopson, & H. Frierson (Eds.), *Continuing the journey to reposition culture and cultural context in evaluation theory and practice* (pp. 91–107). Information Age Publishing.

Greene, J. (2014b). Final reflection. Feminist social inquiry: Relevance, relationships, and responsibility. In S. Brisolara, D. Seigart, & S. SenGupta (Eds.), *Feminist evaluation and research: Theory and practice* (pp. 333–342). The Guilford Press.

Greene, J. C. (2015). The emergence of mixing methods in the field of evaluation. *Qualitative Health Research, 25*(6), 746–750. https://doi.org/10.1177/ 1049732315576499

Greene, J. C. (2016). Advancing equity: Cultivating an evaluation habit. In S. I. Donaldson & R. Picciotto (Eds.), *Evaluation for an equitable society* (pp. 49–65). Information Age Publishing.

Greene, J. C. (2018). In gratitude to Jim and Whit. *New Directions for Evaluation, 2018*(157), 75–77. https://doi.org/10.1002/ev.20291

Greene, J., Boyce, A., & Ahn, J. (2011). *A values-engaged educative approach for evaluating education programs: A guidebook for practice.* Retrieved from https://comm.eval.org/HigherLogic/System/DownloadDocumentfile.ashx? DocumentFileKey=75bc9c3b-b169-4529-b2d3-642056d95f35

Greene, J., & Caracelli, V. (1997a). *Advances in mixed-method evaluation: The challenges and benefits of integrating diverse paradigms.* Jossey-Bass.

Greene, J. C., & Caracelli, V. J. (1997b). Defining and describing the paradigm issue in mixed-method evaluation. *New Directions for Evaluation, 1997*(74), 5–17. https://doi.org/10.1002/ev.1068

Greene, J. C., Caracelli, V. J., & Graham, W. F. (1989). Toward a conceptual framework for mixed-method evaluation designs. *Educational Evaluation and Policy Analysis, 11*(3), 255–274. https://doi.org/10.2307/1163620

Greene, J. C., DeStefano, L., Burgon, H., & Hall, J. (2006). An educative, values-engaged approach to evaluating STEM educational programs. *New Directions for Evaluation, 2006*(109), 53–71. https://doi.org/10.1002/ev.178

Greene, J., Kreider, H., & Mayer, E. (2011). Combining qualitative and quantitative methods in social inquiry. In B. Somekh & C. Lewin (Eds.), *Theory and methods in social research.* (2nd ed., pp. 259–266). SAGE.

Greene, J. C., Lipsey, M. W., Schwandt, T. A., Smith, N. L., & Tharp, R. G. (2007). Method choice: Five discussant commentaries. *New Directions for Evaluation, 2007*(113), 111–127. https://doi.org/10.1002/ev.218

Greene, J. C., Millet, R., & Hopson, R. (2004). Evaluation as a democratizing practice. In M. Braverman, N. Constantine, & J. Slater (Eds.), *Foundations and evaluation: Contexts and practices for effective philanthropy* (pp. 96–118). Jossey-Bass.

Greene, J. C., Sommerfeld, P., & Haight, W. (2010). *Mixing methods in social work research.* In I. Shaw, K. Briar-Lawson, J. Orme, & R. Ruckdeschel (Eds.), The SAGE handbook of social work research (pp. 315–331). SAGE.

Griffith, J. C., & Montrosse-Moorhead, B. (Eds.). (2014). Revisiting truth, beauty, and justice: Evaluating with validity in the 21st century. Wiley Periodicals.

Hall, J., Greene, J., & Ahn, J. (2012). Values engagement in evaluation: Ideas, implications, and illustrations. *American Journal of Evaluation, 33*(2), 195–207. https://doi.org/10.1177/1098214011422592

Hood, S., Hopson, R. K., & Kirkhart, K. E. (2015). Culturally responsive evaluation. In K. E. Newcomer, H. P. Hatry, & J. S. Wholey (Eds.), *Handbook of practical program evaluation* (pp. 281–317). Jossey-Bass. https://doi.org/10.1002/9781119171386.ch12

Hopson, R. K. (Ed.). (2004). How and why language matters in evaluation. *New Directions for Evaluation, 2000*(86), 1–3. https://doi.org/10.1002/ev.1167

Hopson, R. K. (2005). Reinventing evaluation. *Anthropology and Education Quarterly, 36*(3), 289–295.

Hopson, R., & Cram, F. (Eds.). (2018). *Tackling wicked problems in complex ecologies: The role of evaluation.* Stanford University Press.

Hopson, R., Miller, P., & Lovelace, T. (2016). Community-university partnerships as vehicles of radical leadership, service, and change: A critical brokerage perspective. *Leadership and Policy in Schools, 15*(1), 26–44. https://doi.org/10.1080/15700763.2015.1071402

House, E. R. (1980). *Evaluating with validity.* SAGE Publications.

House, E. R., & Howe, K. R. (1999). *Values in evaluation and social research.* SAGE Publications.

Johnson, J., Hall, J., Greene, J. C., & Ahn, J. (2013). Exploring alternative approaches for presenting evaluation results. *American Journal of Evaluation, 34*(4), 486–503. https://doi.org/10.1177/1098214013492995

LaVelle, J. M. (2018). *2018 Directory of Evaluator Education Programs in the United States.* University of Minnesota Libraries Publishing.

LaVelle, J. M. (2020). Educating evaluators 1976–2017: An expanded analysis of university-based evaluation education programs. *American Journal of Evaluation. 41*(4), 494–509. https://doi.org/10.1177%2F1098214019860914

Miller, P., Brown, T., & Hopson, R. (2011) Centering love, hope, and trust in the community: Transformative urban leadership informed by Paulo Freire. *Urban Education, 46*(5), 1078–1099. https://doi.org/10.1177/0042085910395951

O'Shea, J. A. (1985). A Journey to the midway: Ralph Winfred Tyler. *Educational Evaluation and Policy Analysis, 7*(4), 447–459. https://doi.org/10.3102/0162 3737007004447

Smith, N. L, & Brandon, P. R. (2007). *Fundamental issues in evaluation.* Guilford Press.

Tarsilla, M. (2010). Theorists' theories of evaluation: A conversation with Jennifer Greene. *Journal of MultiDisciplinary Evaluation, 6*(13), 209–219.

Williams, D. D. (2018). Twenty-nine evaluation lives. *New Directions for Evaluation, 2018*(157), 1–7. https://doi.org/10.1002/ev.20273

CHAPTER 3

FROM MULTIPLE METHODS TO MIXED METHODS, AND FROM GENERAL PURPOSES TO SPECIFIC DESIGNS

The Greene Light

Melvin M. Mark
The Pennsylvania State University

ABSTRACT

This chapter will explore the pivotal role that Jennifer Greene played in two major transitions in the theory and practice of combining qualitative and quantitative methods. First is the transition from earlier thinking about "multiple methods" to more contemporary conceptualizations of "mixed methods." While this may sound like a simple matter of terminology, the distinction is an important one conceptually and pragmatically. The term *mixed methods* points to a special and distinctive status for combining qualitative and quantitative methods (as opposed to, say, a combination of two quantitative methods). The second transition to be explored in this chapter is the shift

Disrupting Program Evaluation and Mixed Methods Research for a More Just Society, pages 35–47
Copyright © 2023 by Information Age Publishing
www.infoagepub.com

from focusing on a small set of general purposes for combining qualitative and quantitative methods to a set of specific mixed-methods research designs that can be employed in service of one or more of those general purposes (a set of purposes which Jennifer also clarified and expanded).

The title of this book highlights Jennifer Greene's contributions to disrupting program evaluation and mixed-methods research for a more just society. When I was first invited to contribute to the volume, "disrupting" caught my attention. The word "disrupt" shares a common origin with "rupture." Common definitions of disrupt allude to breaking apart, causing upheaval, or throwing into disorder. Of course, the word disrupt can have less negative, indeed even positive connotations, referring to someone or something that interrupts the normal course of events, moving things in a better direction. One stimulus to a more positive sense of the term was a 1997 book that offered the concept of a "disruptive innovation" (Christensen, 1997). Since then, disruption has had a very positive tenor in the innovation space, for example, when you Google "disrupt the sock industry."

Before being invited to contribute a chapter to this volume, I had not thought of Jennifer Greene as a disruptor. In part, I'm old enough—and not enough of a practitioner of business-speak—that the term's negative rather than positive connotations tend to come to my mind. And the terms I would use to describe Jennifer are uniformly positive, personally and professionally. On reflection, however, Jennifer quite rightly is labeled a disruptor. She has interrupted the normal course of events within program evaluation, within mixed-methods research, and especially at their intersection. She is responsible for, or at least has greatly contributed to, multiple disruptive innovations as detailed throughout this book. This includes, but is not limited to, her development of a values-engaged approach to evaluation (e.g., Greene, 1997; Greene et al., 2006) and her furthering of inclusive dialogical processes (Greene, 2001).

In addition to her many and important direct contributions, Jennifer has given what I have come to think of as the "Greene light" for others to proceed, to continue down the road with the disruptions she has initiated. She has stimulated enough traffic down these roads that they have become the normal pathway. In fact, in some cases, those who are newer to these roads may not even realize the disruption that occurred previously or recognize Jennifer's role in it. One aspect of the Greene light is that it signals "go" to a range of perspectives. For example, in closing a chapter based on interviews with early developers of mixed-methods research, Leech (2010) quotes Greene: "Let's keep this conversation [about the field of mixed-methods research] open and dynamic and respectful of the different positions that exist" (p. 269). The Greene light, so to speak, gives the go-ahead to traffic in more than one lane.

In this chapter, I explore the pivotal role that Jennifer Greene played in two major disruptions in the theory and practice of combining qualitative and quantitative methods. First is the transition from earlier thinking about "multiple methods" to more contemporary conceptualizations of "mixed methods." While it may sound like this is a trivial matter of terminology, the distinction is important both conceptually and pragmatically. The term *mixed methods* typically points to a special and distinctive status for combining at least one qualitative method and at least one quantitative method (Greene, 2007; Tashakkori & Teddlie, 1998). In contrast, the term "multiple methods" applies as easily to a combination of two quantitative methods or two qualitative methods, with no special status for mixing qualitative and quantitative (Mark, 2015).

The second transition I explore in this chapter involves shifting from an early focus on a small set of general purposes for combining qualitative and quantitative methods. To be clear, Jennifer and her colleagues did contribute to thinking about the purposes that can be served by mixing quantitative methods. They clarified and expanded the recognized set of purposes (e.g., Greene, 2007; Greene et al., 1989). However, Jennifer and her collaborators went further. They identified specific mixed-method *research designs* that can be employed in service of one or more of those general purposes, as well as the design features that add up to a mixed-method design (Tashakkori & Teddlie, 2003). This shift set into motion a substantial literature on mixed-method research designs, which continues unabated (see, e.g., Billups, 2020).

In exploring these two transitions, this chapter, in a sense, brings qualitative and quantitative together to show how Jennifer's work was foundational. Again, the disruptions resulted, first, in far more attention being given to mixed (than to multiple) methods and, second, in the expansion beyond an earlier focus on general purposes for combining methods to an emphasis on the characteristics (and situationally appropriate selection) of alternative mixed-method designs. The qualitative evidence I present consists of comments drawn from the literature about Jennifer Greene's pivotal role in the disruptions discussed in this chapter. A sampling, rather than a comprehensive cataloging of such comments, is provided. A limited set of relevant publications is also briefly reviewed. The quantitative evidence consists of results from electronic searches conducted using Google Scholar. For example, I provide counts, over time, of how many papers that deal with program evaluation refer to multiple methods versus how many refer to mixed methods. I also examine the pattern of citations of Greene's work to help justify labeling her as a disruptor regarding the two changes I address.

GREENE'S ROLE IN THE MULTIPLE-
MIXED-METHODS TRANSITION

As noted previously, the term *mixed methods* commonly refers to combining qualitative and quantitative methods in a single study or linked series of studies. In contrast, the term *multiple methods* (or multimethod research) covers a broader scope, referring to combining two or more methods, with no implication that both qualitative and quantitative methods are involved. For the most part, it appears that attention to multiple methods occurred before explicit attention to mixed methods. While numerous exemplars of combining methods came earlier, key publications providing a conceptual foundation for multiple methods appeared in the late 1950s and 1960s (e.g., Campbell & Fiske, 1959; Webb et al., 1966). Comparable publications focusing on mixed methods tended to come later (e.g., Jick, 1979; Greene & McClintock, 1985; Greene et al., 1989).

The Mid-to-Late Eighties Disruptive Work

Although Greene's contributions to mixed methods in evaluation and beyond have continued in the years that followed (e.g., Caracelli & Greene, 1993; Greene, 2007; Greene & Caracelli, 1997), the real disruptive work, in my view, appeared between 1985 and 1990. Setting the stage were two 1985 publications, in which Greene and McClintock (1985) and McClintock and Greene (1985) described an "independent, concurrent mixed-methods" evaluation, with the qualitative and quantitative components directed independently by the two authors. Greene and McClintock (1985) discuss the benefits and challenges of achieving triangulation with this mixed-method design. With the benefit of hindsight, it is easy to see how this evaluation helped spur the subsequent insights Jennifer brought to the developing literature on mixed-method evaluations.

The real disruption flowed from the Greene et al.'s (1989) paper. In this article, Greene and her colleagues drew on past frameworks and an inductive analysis of evaluations using mixed methods. Greene et al. (1989) described five possible purposes for mixing methods: triangulation, complementarity, initiation, expansion, and development. These are described in a later section. Greene and her coauthors also described seven design characteristics on which mixed-method designs can vary. In addition to *purpose*, in the 1989 paper, these included *methods, phenomena*, and *paradigms*, all of which can vary in terms of the degree of similarity or difference between the qualitative and quantitative components of the study. The other features described by Greene et al. (1989) are *timing* (the methods are conducted at the same time versus one method is sequenced before the other), *status* or dominance (one method is primary and the other secondary versus the different methods are equal in standing), and *(in)dependence* (the two methods are carried out without one informing the other versus one method and its results inform the other). Various combinations of these characteristics result in different mixed-method designs.

A paper by Caracelli and Greene (1993) falls outside the time frame I have chosen to highlight. And it falls outside of the framework of this chapter. The Caracelli and Greene (1993) paper, in many ways a companion to the Greene et al. (1989) paper, laid important groundwork for another kind of framework. Specifically, the 1993 paper is foundational to frameworks that describe alternative ways of *combining data* from the qualitative and quantitative components of a mixed-methods study (e.g., Brannen & O'Connell, 2015; Onwuegbuzie & Hitchcock, 2015).

Views Regarding Greene's Role in the Transitions

It is easy to find experts in mixed methods who point to Jennifer Greene's key role in founding the field of mixed-methods research, especially her role in establishing this focus within the field of evaluation. For example, in her 2018 book, *Mixed-Methods Design in Evaluation*, Donna Mertens—herself a major contributor to the mixed-methods literature—writes:

> Formal recognition of mixed methods as an important area for discussion in the evaluation community was influenced by Greene and Caracelli's (1997) publication in a volume of *New Directions for Evaluation* on that topic that discussed the role mixed methods can play in evaluation. (p. 3)

Mertens further presents her own book as a kind of continuation of the 1997 Green and Caracelli *NDE*: "This [Greene and Caracelli *NDE*] volume began an important discussion that is expanded on in this [i.e., Mertens's 2018] book" (p. 3).

Recognition of Jennifer's key role in the development of mixed-methods research as an important area is not limited to those working in program evaluation. In a chapter in the *SAGE Handbook of Mixed Methods in Social and Behavioral Research*, Leech (2010) worked with the Handbook editors to identify six "individuals who significantly contributed to the beginning of the field of mixed-methods research" (p. 255). Not surprisingly, Jennifer Greene was one of these. The others were Julia Brannen, Alan Bryman, John Creswell, David Morgan, and Janet Morse. A case can be made that Greene's influence on the interest in mixed methods within program evaluation was strongest in the early days of its development.[1] Creswell (2012) also identified a handful of important figures in the early development of mixed-method research as an area. In addition to three who had appeared on Leech's list (Greene, of course, along with Bryman and Morse), Creswell identified Nigel Fielding, John Hunter, and Allen Brewer. Again, a case can be made that Jennifer Greene has a more central place in evaluation and made the biggest early contribution to the program evaluation's emphasis on mixed methods rather than multiple methods.

Publication Patterns Regarding Mixed and Multiple Methods

Table 3.1 shows the number of publications, by decade, in selected categories over time from 1970 through 2020. One reason for the temporal aggregation and tabular display is that, especially in the earlier years, annual counts are quite low, orders of magnitude smaller than in later years. The information in Table 3.1 was derived from Google Scholar searches in early June 2021. Column A reports the number of articles, books, and other reports (hereafter "items" or "publications") that Google Scholar found which refer to "program evaluation" and "multiple methods." Column B reports the same, but with "mixed methods" instead of multiple methods in the search. Column C presents the percentage of items from Column B that Google Scholar identified that also included "Greene."

The precise counts presented in Tables 3.1 and 3.2 should be viewed with caution. I have not screened all of the items counted. Some publications probably should be screened out (e.g., their mention of program evaluation may have been relatively minor, or they may have referred to the testing of a computer program). Others would not deserve to be screened out conceptually but may be lacking detail. In particular, some of the items are ERIC abstracts that do not include all the information of a full report (and

TABLE 3.1 Number of Publications, by Decade, Referring to Program Evaluation (PE) and Multiple Methods (column A) or to PE and Mixed Methods (column B), and the Percent of Column B Citing Greene (column C)

	Column A: PE and Multiple Methods	Column B: PE and Mixed Methods	Column C: % of Column B Citing Greene
1970–1979	60	5	0.0
1980–1989	320	35	25.7
1990–1999	1,310	197	43.1
2000–2009	3,490	3,330	20.5
2010–2019	5,440	17,300	16.5

the omissions could well include citations to Greene's work, resulting in an underestimate of her influence). Or an ERIC abstract may be duplicative of a subsequent article. Despite these and other limitations, I believe the information in Tables 3.1 and 3.2 suffices for detecting overall trends, especially if one keeps in mind the likely underestimation of Greene's role. Indeed, where they can be compared, the overall patterns in these two tables are consistent with conclusions drawn using a different search approach (see Timans et al., 2019, especially Figure 1 and pp. 201–203).

Table 3.1 shows the expected shift over time from multiple methods to mixed methods. Among publications that include the term "program evaluation," between 1970 and 1989, those referring to multiple methods outnumbered those referring to mixed methods by a ratio of roughly 10 to 1. Items that refer to program evaluation and multiple methods grew substantially over the 50-year period; however, in recent decades, the rate of growth was much higher for items that refer to program evaluation and mixed methods. As a result, in the last decade examined, 2010 to 2019, publications referring to program evaluation and mixed methods were more than three times as common as those referring to program evaluation and multiple methods.

The percentage of items that referred to Greene, as well as to program evaluation and mixed methods, peaked at 43% in the 1990s. The timing of this peak is not surprising, in that Greene and colleagues' major early works on mixed methods, especially the Greene et al. (1989) paper, appeared shortly before that decade. In later periods, authors of a paper involving mixed methods had a wider range of sources to cite. Of course, many of these subsequent sources being cited had been influenced by the work of Greene and her colleagues, especially Greene et al. (1989). Thus, in the later decades, a degree of indirect influence of Greene's work undoubtedly exists. In addition to missing this form of indirect influence, Column C of Table 3.1 likely underestimates the percentage of Google Scholar-found items that refer to both program evaluation and mixed methods and that cite Greene. Some items probably would be excluded from Column B of Table 3.1 if a more rigorous screening process had been used. Other items conceptually belong in Column B but are abstracts and omit citations and references that would occur in a longer document, and so on. Still, the peak percentage suggests that Jennifer was indeed a disrupter, though it suggests she was not the only one in the shift from multiple to mixed methods as a focus of interest.

Mixed Methods: Beyond Purposes to Designs

Although researchers have long been combining methods for a variety of purposes, the early literature on multiple methods initially focused on one use: improving an inference by using more than one method (Tashakkori & Teddlie, 2003). The Campbell and Fiske (1959) article is a key publication in this tradition. Their "multitrait-multimethod matrix" aimed to improve the validity of the measurement of a given trait by triangulating on it using multiple measures. From this and other work, such as Webb and colleagues' (1966) book on unobtrusive measures, in quantitative circles the concept of triangulation is largely associated with Donald Campbell[2] and his associates. Triangulation has also been an important concept in qualitative research, stemming largely from an influential text by Denzin (1978, and later editions). The metaphor of triangulation probably originated in land surveying, where a location of interest can be identified by finding the point of convergence from two known surveying points. Indeed, convergence is a common synonym for triangulation.

In the mid-to-late 1980s, several frameworks were presented that aimed to capture the alternative purposes to which the combining of methods could be put.

An early and influential framework of this kind came from Rossman and Wilson (1985). In short, Rossman and Wilson described three alternative purposes for mixing qualitative and quantitative methods: (a) *corroboration*, that is, a search for triangulation or convergence on an answer; (b) *elaboration*, that is, the enhancement or enrichment of a finding across methods; and (c) *initiation*, that is, the generation of new questions or interpretations. (Mark, 2015, p. 25)

As I have mentioned before,

Mark and Shotland (1987) focused on multiple rather than mixed methods. They described three primary "models" of, or purposes for, multiple methods. One of Mark and Shotland's purposes was *triangulation*, equivalent to Rossman and Wilson's (1985) corroboration. Drawing on Reichardt and Gollob (1987), Mark and Shotland's (1987) second purpose for multiple methods was *bracketing*. The goal of triangulation is the use of different methods to converge on a single answer to a research question, such as finding a single answer about whether and how much a social program helps. Bracketing, in contrast, is based on the idea that two or more methods can provide a *range* of alternative answers. Rather than land surveying and triangulation, a metaphor for bracketing is the confidence interval associated with a statistical test. Triangulation and bracketing each focus on a single research question, with triangulation combining findings from multiple methods to get the best possible single answer, and bracketing focusing on the range of answers across methods. In contrast, Mark and Shotland's (1987) third purpose for multiple methods was to address *complementary purposes*, whereby different methods are employed to address different ends. Mark and Shotland described four kinds of complementary purposes (enhancing interpretability, alternative tasks, alternative levels of analysis, and assessing the plausibility of threats to valid inference). Other variants on complementary purposes could also be described (Mark, 2015, p. 25).

As I have also mentioned,

> Greene and her colleagues (1989), in what has been the most influential of these models, identified five possible purposes for mixed methods. Three of these—*triangulation, complementarity,* and *initiation*—correspond respectively to Rossman and Wilson's (1985) corroboration, elaboration, and initiation. A fourth of Greene et al.'s purposes, *expansion,* refers to the use of different methods for different components of a study. Expansion is similar to Mark and Shotland's (1987) alternative tasks form of complementary purposes. Greene et al.'s fifth purpose is *development.* Development occurs when one method is used to help create (i.e., develop) another method, as when focus groups are carried out to help a researcher create survey items (Mark, 2015, pp. 25–26).

Elsewhere (Mark, 2015), I have attempted to compare and integrate the Rossman and Wilson (1985), Mark and Shotland (1987), and Greene et al. (1989) frameworks describing alternative purposes for combining methods. More important in terms of Jennifer Greene's role as a disruptor is considering how she went beyond these and other discussions about the *purposes* for combining methods, breaking new ground by identifying different kinds of mixed-methods designs. Other subsequent leaders in this domain include Tashakkori and Teddlie (1998, 2003), Creswell (2010), and Creswell et al., 2003). But Jennifer led the way and gave the Greene light to these and others who followed her.

I will not attempt to systematically review or synthesize the alternative frameworks or models that have been proposed to characterize mixed-methods designs (though the authors cited in the previous sentence do). The seminal Greene et al. (1989) approach was sketched earlier. It informed Greene's (2007) later book on mixed-methods research. By way of partial summary, a limited number of dimensions are described to capture key characteristics along which the qualitative and quantitative components of mixed-methods studies can differ. These include *purpose*—whether the two (or more) methods are being used for triangulation, or for complementarity, or for another of purposes that Greene and colleagues (1989) mention—in addition to *timing, status, (in)dependence,* and the *methods themselves* in terms of the degree of similarity or dissimilarity in the two methods.

Publication Patterns Regarding Mixed-Methods Designs

Column A in Table 3.2 reports, by decade, the number of Google Scholar items that include both "program evaluation" and "mixed-method design."

TABLE 3.2 Number of Publications, by Decade, Referring to Program Evaluation and Mixed-Method Design (column A), and the Number (column B) and the Percent (column B) of These Publications Citing Greene (column C)

	Column A: PE and Mixed-Method Design	Column B: Number of Column A Citing Greene	Column C: % of Column A Citing Greene
1970–1979	0	0	—
1980–1989	4	3	75.0
1990–1999	37	33	89.2
2000–2009	194	107	54.1
2010–2019	436	142	32.6

Columns B and C give the number and percentage, respectively, of these items that cite Greene. The number of items referring to mixed-method design, not surprisingly, is substantially less than the number shown in Table 3.1 referring to mixed methods more generally. The absolute number in Table 3.2 may be an underestimate, because I did not use the various phrasings that authors might use to refer to mixed-methods designs. Regardless, there is a sizable growth over time in publications dealing with this topic, increasing 100-fold from a small base in the 1980s to the 2010s.

Greene was cited in three of the four papers in the 1980s that involved program evaluation and mixed-methods designs; she was an author of these three papers. In the 1990s, she was cited in nearly 90% of the items. Perusal of those items that did not include a Greene citation revealed that her influence that decade was even stronger than the 90% citation rate would suggest. Of the four papers not citing Greene, one was a piece by Lois-ellin Datta in the 1997 *NDE* issue that Jennifer co-edited with Caracelli, and in a footnote, Datta thanked Jennifer for "personal communication" (p. 33). Another was an ERIC abstract without full text or references. Given the abstract, it seems more likely than not that Greene was cited in the full paper. A third was a dissertation, and only part of the dissertation was available in the document listed in Google Scholar. The dissertation reported a meta-evaluation of an evaluative review carried out by a state education agency. It is unclear whether Greene was cited in the full document or in one or more of the evaluations on which the meta-evaluation was based. The final item without a reference to Greene was a directory of projects funded by an office within the Department of Education. Again, it is unclear whether an original project that refers to mixed-method designs included citations to Greene in its proposal or reports. And in any case, a directory of projects might be removed from consideration if a more formal screening of all the items had been conducted.

Regardless of whether the rate of citations to Greene in the 1990s papers involving program evaluation and mixed-methods designs is 90% or 100% or somewhere in between, Jennifer's strong impact in this area in the 1990s is clear. With her coauthors, Jennifer appears to be *the* disruptor who expanded beyond previous discussions of the purposes of mixing methods, enlarging attention to a focus on mixed-methods designs and design features.

The proportion of citations to Jennifer's work in mixed-methods design papers in program evaluation declined in each of the next 2 decades. In terms of decade-to-decade change, the rate of citations to Greene drops about 40% with each decade after the 1990s. As Table 3.2 shows, the absolute number of citations to Greene increases across decades, but the proportion of evaluation papers that both refer to mixed-method designs and cite Greene declines. Such is the nature of influence. Even the influence of a key disrupter increasingly becomes indirect and mediated by the work of others over time—especially when that disruptor shifts much of her focus to other areas, as illustrated in the other chapters in this volume.

CONCLUSION

The editors of this volume quite appropriately have portrayed Jennifer Greene's contributions as disruptive. I am grateful that their framing of this book has helped me think about just how disruptive, in the positive sense of that term, Jennifer has been, and in multiple domains of scholarship and practice. In this chapter, I have focused on Jennifer's contributions to two disruptions. First is the shift from multiple methods to mixed methods. Comments from other experts in mixed methods and an analysis of publication counts and citations show Jennifer as key to this shift within evaluation, though not solely responsible for this disruption. A second disruption is the shift from an earlier focus only on general purposes for combining methods to a subsequent focus on mixed-method research designs. In this case, citation analyses point to Jennifer and her colleagues as *the* disrupters, as indicated especially by their near-universal citation in work alluding to mixed-method designs in evaluation during the 1990s. While the number of evaluation papers involving mixed-method designs has grown substantially since the 1990s, the proportion of such papers that cite Green has declined over time. This is one example of the Greene light, whereby others have continued down the road, inspired, enabled, and encouraged by Jennifer's work. And those who have been directly influenced by Jennifer's work help shape the path forward. They influence the conceptualizations and practices of others who follow.

And the Greene light shines on.

NOTES

1. Greene is far more cited in works involving program evaluation than are Brannen, Bryson, or Morse, especially for the fertile period of 1985–1990. Creswell and Morgan are highly cited in works involving program evaluation, but relatively less so in that early period and largely with respect to other topics (focus groups in Morgan's case and faculty evaluation in Creswell's).
2. Those who cite Campbell while advocating confident inferences about a program based on a single randomized trial might well remember this aspect of his work.

REFERENCES

Billups, F. D. (2020). *Qualitative data collection tools: Design, development, and applications.* Sage.

Brannen, J., & O'Connell, R. (2015). Data analysis I: Overview. In S. Hesse-Biber & R. B. Johnson (Eds.), *The Oxford handbook of multimethod mixed-methods research inquiry* (pp. 257–274). Oxford University Press.

Campbell, D. T., & Fiske, D. W. (1959). Convergent and discriminant validation by the multitrait–multimethod matrix. *Psychological Bulletin, 56*(2), 81–105. https://doi.org/10.1037/h0046016

Caracelli, V. J., & Greene, J. C. (1993). Data analysis strategies for mixed-method evaluation designs. *Educational Evaluation and Policy Analysis, 15*(2), 195–207. https://doi.org/10.3102/01623737015002195

Christensen, C. M. (1997). *The innovator's dilemma: When new technologies cause great firms to fail.* Harvard Business School Press.

Creswell, J. W. (2010). Mapping the developing landscape of mixed methods research. In A. Tashakkori & C. Teddlie (Eds.), *SAGE handbook of mixed methods in social & behavioral research* (2nd ed., pp. 45–68). SAGE.

Creswell, J. (2012). *Qualitative inquiry and research design: Choosing among five approaches.* SAGE Publications.

Creswell, J. W., Plano Clark, V. L., Gutmann, M. L., & Hanson, W. E. (2003). Advanced mixed methods research designs. In A. Tashakkori & C. Teddlie (Eds.), *Handbook of mixed methods in social & behavioral research* (pp. 209–240). SAGE Publications.

Datta, L.-e. (1997), A pragmatic basis for mixed-method designs. *New Directions for Evaluation, 1997,* 33–46. https://doi.org/10.1002/ev.1070

Denzin, N. K. (1978). *The research act: A theoretical introduction to sociological methods.* McGraw-Hill.

Greene, J. C. (1997). Evaluation as advocacy. *Evaluation practice, 18*(1), 25–35. https://doi.org/10.1177/109821409701800103

Greene, J. C. (2001). Dialogue in evaluation: A relational perspective. *Evaluation, 7*(2), 181–187. https://doi.org/10.1177/135638900100700203

Greene, J. C. (2007). *Mixed methods in social inquiry.* John Wiley & Sons.

Greene, J. C., & Caracelli, V. (Eds.). (1997). *Advances in mixed-method evaluation: The challenges and benefits of integrating diverse paradigms.* Jossey-Bass.

Greene, J. C., Caracelli, V., & Graham, W. (1989). Toward a conceptual framework for mixed-method evaluation designs. *Educational Evaluation and Policy Analysis, 11*(3), 255–274. https://doi.org/10.3102/01623737011003255

Greene, J. C., DeStefano, L., Burgon, H., & Hall, J. (2006). An educative, values-engaged approach to evaluating STEM educational programs. *New Directions for Evaluation, 2006*(109), 53–71. https://doi.org/10.1002/ev.178

Greene, J., & McClintock, C. (1985). Triangulation in evaluation: Design and analysis issues. *Evaluation Review, 9*(5), 523–545. https://doi.org/10.1177/0193841X8500900501

Jick, T. D. (1979). Mixing qualitative and quantitative methods: Triangulation in action. *Administrative Science Quarterly, 24*(4), 602–611. https://doi.org/10.2307/2392366

Leech, N. (2010). Interviews with the early developers of mixed methods research. In A. M. Tashakkori & C. B. Teddlie (Eds.), *SAGE handbook of mixed methods in social & behavioral research* (pp. 253–272). SAGE.

Mark, M. M. (2015). Mixed and multimethods in predominantly quantitative methods, especially experiments and quasi-experiments. In S. Hesse-Biber & R. B. Johnson (Eds.), *The Oxford handbook of multimethod mixed methods research inquiry* (pp. 21–41). Oxford University Press.

Mark, M. M., & Shotland, R. L. (1987). Alternative models for the use of multiple methods. In M. M. Mark & R. L. Shotland (Eds.), *Multiple methods in program evaluation* (pp. 95–100). Jossey-Bass.

McClintock, C., & Greene, J. (1985). Triangulation in practice. *Evaluation and Program Planning, 8*(4), 351–357. https://doi.org/10.1016/0149-7189(85)90032-1

Mertens, D. (2018). *Mixed methods design in evaluation.* SAGE.

Onwuegbuzie, A. J., & Hitchcock, J. (2015). Advanced mixed analysis approaches. In S. Hesse-Biber & R. B. Johnson (Eds.), *The Oxford handbook of multimethod mixed methods research inquiry* (pp. 275–295). Oxford University Press.

Reichardt, C. S., & Gollob, H. F. (1987). Taking uncertainty into account when estimating effects. In M. M. Mark & R. L. Shotland (Eds.), *Multiple methods in program evaluation* (pp. 7–22). Jossey-Bass.

Rossman, G. B., & Wilson, B. L. (1985). Numbers and words: Combining quantitative and qualitative methods in a single large-scale evaluation study. *Evaluation Review, 9*(5), 627–643. https://doi.org/10.1177/0193841X8500900505

Tashakkori, A., & Teddlie, C. (1998). *Mixed methodology: Combining qualitative and quantitative approaches.* SAGE Publications.

Tashakkori, A., & Teddlie, C. (2003). The past and the future of mixed model research: From "methodological triangulation" to "mixed model designs." In A. Tashakkori & C. Teddlie (Eds.), *Handbook of mixed methods in social & behavioral research* (pp. 671–702). SAGE Publications.

Timans, R., Wouters, P., & Heilbron, J. (2019). Mixed methods research: What it is and what it could be. *Theory and Society, 48*(1), 193–216. https://doi.org/10.1007/s11186-019-09345-5

Webb, E. J., Campbell, D. T., Schwartz, R. D., & Sechrest, L. (1966). *Unobtrusive measures: Nonreactive research in the social sciences.* Rand-McNally.

CHAPTER 4

THE INTERPRETIVIST-HUMANIST EVALUATOR

Thomas A. Schwandt
University of Illinois at Urbana–Champaign

ABSTRACT

In her evaluation scholarship and practice, Jennifer Greene exhibits in an exemplary way the orientation of an interpretivist-humanist. Her work builds upon a tradition of qualitative inquiry in evaluation, and she complements that by broadening and developing insights into the democratic aims of evaluation. In her practice, speeches, and writing, she displays an acute sensitivity to the perspectives and values of evaluators and the stakeholders with whom they work and promotes norms of equity and social justice.

Jennifer Greene and I have been friends and colleagues for more than 35 years, including 16 or so we spent together in the Department of Educational Psychology at the University of Illinois. I have had many opportunities to discuss evaluation theory and practice with Jennifer, and I have witnessed her give conference papers and invited speeches, teach graduate students, and serve as an involved and committed citizen of our department. Given this long association with Jennifer, I had initially proposed to

Disrupting Program Evaluation and Mixed Methods Research for a More Just Society, pages 49–58
Copyright © 2023 by Information Age Publishing
www.infoagepub.com
49

focus on Jennifer as an exemplar of the scholar-practitioner. Scholar-practitioners are individuals who "have one foot each in the worlds of academia and practice and are pointedly interested in advancing the causes of both theory and practice" (Tenkasi & Hay, 2008, p. 50). As a researcher, teacher, and practitioner, Jennifer seamlessly weaved together theory and practice. Some years ago, Charles McClintock (2004), a former colleague of Jennifer's at Cornell University, wrote that

> The term scholar practitioner expresses an ideal of professional excellence grounded in theory and research, informed by experiential knowledge, and motivated by personal values, political commitments, and ethical conduct. Scholar practitioners are committed to the well-being of clients and colleagues, to learning new ways of being effective, and to conceptualizing their work in relation to broader organizational, community, political, and cultural contexts. (p. 393)

This description is an accurate characterization of Jennifer in her roles as educator, practicing evaluator, and university professor. Yet, I suspected that commenting solely on her role as scholar-practitioner might be redundant since I anticipated that other chapters authored by former students that Jennifer mentored and advised would offer testimony to that effect. Thus, I take a somewhat different approach in this chapter and focus on how Jennifer exhibits in her scholarship and practice an interpretivist-humanist orientation. An orientation to one's work is different than a role. We occupy multiple roles in society as parents, grandparents, friends, colleagues, teachers, students, and professionals of all kinds. A role connotes a function one plays in certain situations; it involves a set of behaviors, norms, and obligations. By using the term *orientation* rather than *role*, I aim to convey a set of attitudes, dispositions, and perspectives that are more like traits, virtues of character, or temperaments than functions or duties. The interpretivist-humanist orientation informs the way Jennifer understands what is expected of evaluators by clients and the public, as well as the kind of knowledge that informs good evaluation practice. For Jennifer, that knowledge is a type of understanding arising from meaningfully engaging with others and from examining the values one holds and promotes as a professional.

THE INTERPRETIVIST

Interpretivists in social science and evaluation focus on meaning—on understanding the beliefs, motives, values, perspectives, and reasons of social actors to understand social reality. They typically employ qualitative methods to both access and render accounts of meaning. Jennifer's work on mixed methods (see Chapter 3, p. 50) reveals that she does not deny the value of quantitative methods. Yet, her methodological stance reflects strong commitments to the qualitative, interpretivist way of practicing evaluative inquiry (Greene, 1994, 1996). This orientation grows out of the influences of Egon Guba's work (Greene, 2008) and a commitment to a responsive approach to evaluation as advocated by Bob Stake. Early in her career, Jennifer authored entries on "naturalistic interviewing," "naturalistic data collection," and "qualitative data analysis and interpretation" in an edited book on resources for teaching evaluation (Mertens, 1989). In much of her writing based on empirical evaluation studies, Jennifer makes use of cases and offers accounts of contextualized meaning (Greene 2000; Greene et al., 2001; Greene, 2006; Hall et al., 2012). Both are hallmarks of an interpretive orientation to social inquiry.

Responsiveness

An interpretive orientation involves being responsive to the views of those involved with an evaluation. It demands that the evaluator be approachable, open-minded, and attentive to perspectives and values other than one's own. For example, in an interview with Jody Fitzpatrick (2001) about an evaluation that Jennifer and colleagues conducted of the W. K. Kellogg Natural Resources Leadership Program, Jennifer explained:

> My approach to evaluation is responsive, in Bob Stake's sense of the word, in that the evaluation tries to become attuned to the issues and concerns of people in the context of this program. And these issues are always ones about values. Second, my methodology is inclusive in that multiple, diverse perspectives, stances, and concerns are included in the evaluation. (p. 85)

Being an evaluator is a role; being responsive as an evaluator is an orientation. In her autobiographical reflection, "Making the World a Better Place Through Evaluation" (Greene, 2013), where she discussed important influences on her work, Jennifer reiterated this indebtedness to the idea of being responsive to the interests, perspectives, and issues of parties to an evaluation.

In describing their "fourth-generation evaluation" approach, fellow qualitative and interpretivist evaluators Guba and Lincoln (1989) advocated the use of a dialogic, iterative, and dialectic process in which evaluators surfaced and then made plain to all stakeholders the range of stakeholder issues and inherent conflicts in values and perspectives. Concurring with Guba and Lincoln's views, Jennifer claims that a responsive perspective on evaluation "situates evaluation as a process wherein the evaluator's role is not only to describe various values and issues but also to create opportunities for diverse stakeholder views to be heard by others" (Hall et al., 2012, p. 196). She added that "the purpose of evaluative dialogue is to enable stakeholders to more deeply understand and respect, although not necessarily agree with, one another's perspectives" (Greene, 2001, p. 182), and that a focus on inclusion of multiple viewpoints "is intended to acknowledge and respect the plurality of stakeholder perspectives and experiences present in the context, rather than to validate them" (Hall et al., 2012, p. 198).

Like both Guba and Lincoln, as well as Stake, Jennifer combines methodological and ethical rationales for the idea of inclusion of stakeholder interests, values, and perspectives. Yet, her view of responsiveness extends beyond their views when she weds the focus on inclusion to an explicit political commitment of the evaluator to promote equity—understood as the fair and just treatment of program stakeholders, specifically, how well the evaluand affords program access, meaningful participation, and accomplishment for all relevant stakeholders (Hall et al., 2012, p. 198).

THE HUMANIST

The terms *humanist* and *humanism* acquire different meanings depending on their interpretations in several scholarly traditions. Here, I appropriate an understanding of the terms found in the literature in medicine and medical education that speaks to the character of a particular kind of medical professional (Lee Roze des Ordons et al., 2018; Miller & Schmidt, 1999). There one finds the notions of *humanistic care* and *humanistic engagement*. Both refer to a way of being and "denote a set of deep-seated personal convictions about one's obligations to others, especially those in need" (Cohen, 2007, p. 1029). Characteristics of humanistic engagement include sensitivity to cultural backgrounds and values, attunement to the preferences of others, perspective taking, recognition of the shared elements of the human condition, and a relational focus. As defined here, a humanistic orientation is a professional ethic, and Jennifer specifically linked her interest in democratic evaluation (as discussed below) to an ethic of care (Simons & Greene, 2018).

In Jennifer's work, humanistic engagement as a set of commitments to others and a constellation of personal dispositions has three distinguishing features: It is manifest as values-engagement, it is a requirement for an educative approach to evaluation that serves democratic pluralism, and it animates evaluation professionalism.

Values-Engagement

Values-engagement has a twofold focus. First, it means the explicit consideration of how values are at work in shaping the purpose, framing, and conduct of evaluation. In a chapter she prepared for a UNICEF-sponsored collection of papers on equitable evaluation, Jennifer explained:

> The term "values-engagement" ... signal[s] explicit attention to values as part of the evaluation process and to the central role that values play in our evaluation practice. From the framing of evaluation questions to the development of an evaluation design and methods, and from the interactions of stakeholders in the evaluation process to the especially important task of making judgments of programme quality ... Engagement thereby suggests a kind of quiet insistence that questions of value be addressed throughout the evaluation, at every turn and every decision point—so values become interlaced with, knitted and knotted within evaluative thinking and judging. This is our aspiration. (Greene, 2012, p. 198)

Second, values-engagement signifies the personal commitment of the evaluator to describe and engage "the perspectives, concerns, and values of all legitimate stakeholders in the evaluation, with particular attention to ensuring inclusion of the interests, perspectives, and values of those traditionally unheard, underrepresented, or least well served in that context" (Hall et al., 2012, p. 198). Jennifer views this humanistic engagement with stakeholders in terms of what she calls a "relational perspective" made possible through dialogue:

> Dialogue in evaluation ... is a fundamental value commitment to engagement, particularly engagement with the relational, moral and political dimensions of our contexts and craft ... [E]ffective dialogue in evaluation is inclusive of all legitimate interests and involves interactions that are respectful, reciprocal and equitable. (Greene, 2001, p. 181)

Educative Approach to Evaluation Serving Democratic Decision-Making

Greene regards this attention to the inclusion of multiple perspectives of stakeholders in evaluation not only as a professional ethical (and methodological) commitment but as a requirement for an educative approach to evaluation. Here she reflects (Greene, 2013) influences of her major professor, Lee J. Cronbach, who famously published "Ninety-Five Theses for Reforming Program Evaluation" (Cronbach & Associates, 1980) that captured key features of his theory of evaluation as an educative undertaking that enlightens all participants in a democratic, pluralistic process of decision-making. Four of these theses shown below are particularly significant for understanding how Jennifer views humanistic engagement as a requirement for an educative approach to evaluation that serves democratic pluralism.

Thesis 1: Program evaluation is a process by which society learns about itself.

Thesis 2: Program evaluation should contribute to enlightened discussion of alternative plans for social action.

Thesis 11: A theory of evaluation must be as much a theory of political interaction as it is a theory of how to determine facts.

Thesis 93: The evaluator is an educator; his [sic] success is to be judged by what others learn.

As a political philosophy, democratic pluralism involves recognizing and affirming a diversity of interests, convictions, and cherished values within the body politic. While to the best of my knowledge, Jennifer does not discuss to any extent the companion notion of value pluralism in her work—the notion that there are multiple values that may be equally correct and important yet in conflict with one another—I infer from many of her writings that she would endorse this idea as well. The philosopher Isaiah Berlin, a strong defender of pluralism captured the spirit of the pluralist doctrine and its close affiliation with the idea of humanistic engagement in this observation:

> Let us have the courage of our admitted ignorance, of our doubts and uncertainties. At least we can try to discover what others... require, by taking off the spectacles of tradition, prejudice, dogma, and making it possible for ourselves to know men [sic] as they truly are, by listening to them carefully and sympathetically, and understanding them and their lives and their needs. (Berlin 1953/1994, p. 258)

Jennifer promotes evaluation as having a primary role in contributing to democratic decision-making (Greene, 1997, 2001, 2013). Evaluators can fulfill that role only if, in their practice, they are committed to

> full participation of all legitimate stakeholder interests in the conversation and in all relevant decisions about a particular program's merit and worth, with democratic principles of equality, fairness, and justice as guides to both the conversation and the decision-making. This commitment to democratic pluralism constitutes the regulative ideal for this vision of evaluation, its essential value commitment. (Greene, 1997, p. 28)

And as noted in Chapter 3, Jennifer's stance on the importance of mixed methods is highly consistent with this view. Mixed-methods strategies require a kind of reflexive engagement with one's practice that rests on an openness to diversity, acceptance of difference, and tolerance of ambiguity (Greene et al., 2001).

Evaluation Professionalism

In writing about humanistic engagement in medicine, Cohen (2007) argues that it "provides the passion that animates professionalism" (p. 1029). Professionalism is not simply about the attitudes and actions of individuals but also about the ethos of the profession (Schwandt, 2017). The notion of an ethos captures the sum of a professional group's moral principles, core values, epistemic and aesthetic dispositions, and aspirations that each member of the group takes into consideration in interacting with others in a professional context.

For Jennifer, the ethos of evaluation professionalism stimulated by the idea of humanistic engagement should be evident in evaluators acting as "public scientists" who actively contribute to discussion and actions about public issues (Greene, 1997). Here again, we hear echoes of Cronbach's theses as listed above. She argues that evaluators should have a strong voice in contemporary conversations about social problems and policies and society's challenges. She claims that evaluators should assume the mantel of "scientific citizenship" (Greene, 1996), thereby taking social responsibility for the political, moral, and value consequences of their work. She made this obligation particularly explicit in discussing how evaluators, and all social inquirers more generally, need to address their "sociopolitical commitments" by asking questions such as these:

> Whose interests should be served by this approach to social inquiry, and why? Where is this inquiry located in society? Does the study contribute to collective theoretical knowledge; is it a "knowledge producer"? Does it advise governmental decision makers? Is the study located in a protected space, separate and apart from the political fray? Or is it located in the midst of contestation, in a position of social critique or advocacy for particular interests and positions? (Greene, 2006, p. 94)

FINAL THOUGHTS

The title of this volume points to Jennifer as a "disrupter" of program evaluation. A disrupter causes a disturbance or problem; throws a state of affairs into confusion; creates a kerfuffle, commotion or ruckus; or drastically alters the structure of something. I take exception to that characterization. Jennifer said it best in claiming that "engagement [in evaluation] . . . suggests a kind of quiet insistence that questions of value be addressed" (Greene 2012, p. 198). In her character, practice, and scholarship, Jennifer does not disrupt but seeks to persuade.

Based on my long association with Jennifer and an examination of her work, I also regard her as a steward of the discipline to borrow a phrase introduced several years ago during the Carnegie Initiative on the Doctorate sponsored by the Carnegie Foundation for Advancement of Teaching. Golde (2006) emphasizes that the label *steward* conveys a role that transcends a collection of accomplishments and skills. It has an ethical and moral dimension (p. 12). According to Perry (2016), the steward simultaneously occupies places in the worlds of practice and the world of scholarship (p. 300). In the former, the steward is a professional, trained in the applied skills of the profession to which she belongs. She adheres to professional standards and codes and acts on behalf of that profession. In the world of scholarship, she is a scholarly practitioner using practical research and applied theory as tools for change. The steward is also both a curator and cultivator of the discipline. Recognizing, on the one hand, an indebtedness to traditions of thought and ideas that precede her work, the steward simultaneously aims to promote new ways of thinking to extend the discipline's theory and practice. In acting in this way, the steward adopts a sense of purpose larger than oneself (Golde, 2006).

As a professional, the steward is also humble about the value of the expertise she brings to a task and about expectations for her contribution to improving a situation. In a presentation to the Research on Evaluation Special Interest Group of the American Educational Research Association on how evaluation practice might serve the aims of securing social justice, Melvin Hall (2019), a longtime colleague of Jennifer's, claimed that evaluators should exhibit the following attitudes and traits:

- We should not oversell our role and contribution.
- We must be clear about what is left to evaluator disposition and what is normative for the profession.
- We must not overcommit based upon the training and experience evaluators possess.
- We must invent new ways to free evaluation from its institutional limits or acknowledge them openly.
- We must foreground inquiry skills over judgments rendered.

Melvin might well have had Jennifer's scholarship and practice in mind in developing this normative perspective.

REFERENCES

Berlin, I. (1994). *Russian thinkers.* Penguin Books. (Original work published 1953)

Cohen, J. J. (2007). Viewpoint: Linking professionalism to humanism: What it means, why it matters. *Academic Medicine, 82*(11), 1029–1032. https://doi .org/10.1097/01.ACM.0000285307.17430.74

Cronbach, L. J., & Associates. (1980). *Toward reform of program evaluation.* Jossey-Bass.

Fitzpatrick, J. (2001). A conversation with Jennifer Greene. *American Journal of Evaluation, 22*(1), 85–96.

Golde, C. M. (2006.) Preparing stewards of the discipline. In C. M. Golde & G. E. Walker (Eds.), *Envisioning the future of doctoral education* (pp. 3–20). Jossey-Bass.

Greene, J. C. (1994). Qualitative program evaluation: Practice and promise. In N. K. Denzin & Y. S. Lincoln (Eds.), *Handbook of qualitative research* (pp. 530–544). SAGE Publications.

Greene, J. C. (1996). Qualitative evaluation and scientific citizenship: Reflections and refractions. *Evaluation, 2*(3), 277–289. https://doi.org/10.1177/135638 909600200303

Greene, J. C. (1997). Evaluation as advocacy. *American Journal of Evaluation, 18*(1), 25–35. https://doi.org/10.1177/109821409701800103

Greene, J. C. (2000). Challenges in practicing deliberative democratic evaluation. In K. E. Ryan & L. DeStefano (Eds.), *Evaluation as democratic process: Promoting inclusion, dialogue, and deliberation* (pp. 13–26). Jossey-Bass

Greene, J. C. (2001). Dialogue in evaluation: A relational perspective. *Evaluation, 7*(2), 181–187. https://doi.org/10.1177/135638900100700203

Greene, J. C. (2006). Toward a methodology of mixed methods social inquiry. *Research in the Schools Mid-South Educational Research Association, 13*(1), 93–99.

Greene, J. C. (2008). In tribute to Egon Guba. *Qualitative Inquiry, 14*(8), 1360–1365. https://doi.org/10.1177/1077800408325259

Greene, J. C. (2012). Values-engaged evaluation. In Marco Segone (Ed.), *Evaluation for equitable development results* (pp. 192–206). Retrieved from https://www .yumpu.com/en/document/view/47769876/evaluation-for-equitable -development-results-capacityorg

Greene, J. C. (2013). Making the world a better place through evaluation. In M.C. Alkin (Ed.), *Evaluation roots* (2nd ed., pp. 208–217). SAGE Publications.

Greene, J. C., Benjamin, L., & Goodyear, L. (2001). The merits of mixing methods in evaluation. *Evaluation, 7*(1), 25–44. https://doi.org/10.1177/1356 3890122209504

Guba, E. G., & Lincoln, Y. S. (1989). *Fourth generation evaluation.* SAGE Publications.

Hall, J. N., Greene, J. C., & Ahn, J. (2012). Values-engagement in evaluation: Ideas, illustrations, and implications. *American Journal of Evaluation, 33*(2), 195–207. https://doi.org/10.1177/1098214011422592

Hall, M. (2019, April). *Program evaluation and social justice* [Presentation]. Research on Evaluation Special Interest Group at the annual meeting of American Educational Research Association, Toronto, Canada.

Lee Roze des Ordons, A., de Groot, J. M., Rosenal, T., Viceer, N., & Nixon, L. (2018). How clinicians integrate humanism in their clinical workplace—"Just trying to put myself in their human being shoes." *Perspectives on Medical Education,* 7(5), 318–324. https://doi.org/10.1007/s40037-018-0455-4

McClintock, C. (2004). Scholar practitioner model. In A. Distefano, K. E. Rudestam, & R. J. Silverman (Eds.), *Encyclopedia of distributed learning* (pp. 393–396). SAGE Publications.

Mertens, D. (Ed.). (1989). *Creative ideas for teaching evaluation: Activities, assignments, and resources.* Kluwer Nijhoff

Miller, S. Z., & Schmidt, H. J. (1999). The habit of humanism: A framework for making humanistic care a reflexie clinical skill. *Academic Medicine, 74*(7), 800–803. https://doi.org/10.1097/00001888-199907000-00014

Perry, J. A. (2016). The scholarly practitioner as steward of the practice. In V. A. Story & K. A. Hesbol (Eds.), *Contemporary approaches to dissertation development and research methods* (pp. 300–313). Information Science Reference.

Schwandt, T. A. (2017). Professionalization, ethics, and fidelity to an evaluation ethos. *American Journal of Evaluation, 38*(4), 546–553. https://doi.org/10.1177/1098214017728578

Simons, H., & Greene, J. C. (2018). Democratic evaluation and care ethics. In M. Visse & T. Abma (Eds.), *Evaluation for a caring society* (pp. 83–104). Information Age Publishing.

Tenkasi, R. V., & Hay, G. W. (2008). Following the second legacy of Aristotle: The scholar-practitioner as an epistemic technician. In A. B. Shani, S. A. Mohrman, W. A. Pasmore, B. Stymne, & N. Adler (Eds.), *Handbook of collaborative management research* (pp. 49–72). SAGE Publications.

SECTION I

DEMOCRATIZING INQUIRY: FOSTERING INCLUSION, COLLABORATION, AND LEARNING

Section I of the book focuses on how Greene's democratic orientation—emphasizing inclusive, collaborative, culturally responsive, and educative inquiry stances and practices—has disrupted social science inquiry. In Chapter 5, Emily Gates, Priya La Londe, Rebecca Teasdale, and Carolina Standen explore Greene's democratic vision for mixed methods and evaluation, providing examples of how Greene disrupts traditional evaluation practices. In Chapter 6, Veronica G. Thomas discusses Greene's work within the broader context of the call for a more inclusive and socially just evaluation practice, exemplifying how Greene is a "gentle" disruptor. Sharon Brisolara, Tristi Nichols, Kathryn A. Sielbeck-Mathes, and Denise Seigart, in Chapter 7, offer a thoughtful reflection on how Greene's authentic care, communication style, and relationship-building techniques demonstrate her democratic and equity values. The section concludes with Chapter 8 by Abma Tineke and Susan Woelders. This chapter builds on Greene's notion of democratic evaluation to discuss epistemic justice; that is, the notion that all stakeholders should be enabled to participate in the evaluation, particularly those who are marginalized.

CHAPTER 5

DEMOCRATIC IDEALS IN RESEARCH AND EVALUATION

Purposes, Puzzles, and the Public Good

Emily F. Gates
Boston College

Priya Goel La Londe
The University of Hong Kong, Hong Kong S. A. R.

Rebecca M. Teasdale
University of Illinois at Chicago

Carolina Hidalgo Standen
University of La Frontera, Chile

ABSTRACT

As a scholar, evaluator, and mentor, Jennifer C. Greene embodies a democratic vision for mixed-methods and evaluative inquiry. In this chapter, four former students, currently university-based faculty, reflect on Greene's vision

Disrupting Program Evaluation and Mixed Methods Research for a More Just Society, pages 61–75
Copyright © 2023 by Information Age Publishing
www.infoagepub.com
All rights of reproduction in any form reserved.

and practice of democratizing inquiry. This chapter examines three ways she disrupted long-standing ideals in social investigation and pursued alternatives: crafting clear and meaningful *purposes* to guide our work, embracing inquiry as *puzzles* by welcoming the unexpected and divergent, and serving the *public good*. We explore these three ideals in her scholarship, her teaching, and our experiences with her. We close with personal reflections and aspirations to forward her contributions.

Recently, a student asked me (La Londe), "Is the Greene reading for this week important?" While I (La Londe) wanted to make sure the student fully understood the depth and value of Greene's scholarship, I did not want the student to feel intimidated. Channeling Greene's ethic of care, I responded,

> Basically, anything by Jennifer Greene is essential because it will help you think about how you understand research, which is what will help you make sense of how to do research. So, take these readings slowly, and be sure to jot down how these readings shape what you think about diversity, dialogue, and democracy.

I shared this anecdote with the other authors. We all smiled and discussed our shared experiences with trying to capture the complexity and depth of Greene's work on social inquiry. As early-career scholars who teach Greene's scholarship in our classrooms, we are fortunate to have "grown up" under the influences of her mentorship.

This chapter explores contributions to democratizing inquiry through three core ideals and ways these show up in her scholarship, courses, and mentorship. The chapter relates to the section, "Democratizing Inquiry: Fostering Inclusion, Collaboration, and Learning," by exploring how Greene envisions and enacts democratic forms of program evaluation and mixed-methods research. The chapter begins with a grounding in some of the democratic roots and visions Greene draws on. Then, we discuss three core ways Greene disrupts long-standing ideals and democratizes inquiry:

- crafting clear and meaningful *purposes* to guide our work,
- embracing inquiry into empirical *puzzles* by welcoming the unexpected and divergent, and
- envisioning and enacting one's professional responsibility to serve the *public good*.

We explore how these ideals challenge prior ways of working and offer inspiration and practical guidance into democratizing inquiry. We then turn to examples of these ideals in Greene's teaching by sharing highlights from our experiences in her courses. We close with personal reflections and aspirations to forward her contributions.

DEMOCRATIC ROOTS AND VISIONS

Democratic traditions see social inquiry, research, and evaluation as opportunities for democracy in action. Greene (2002, 2005, 2006a) acknowledges the vital roles research plays in representative and deliberative notions of democracy. In "Evaluation, Democracy, and Social Change," Greene (2006a) reviews Barry MacDonald's representative democratic approach; Ernest House and Kenneth Howe's deliberative democratic approach; and participatory, critical, culturally and contextually responsive, and indigenous approaches to evaluations. Here, Greene advances a third role for democracy in inquiry, research, and evaluation, which envisions democracy as social practice and evaluation as public craft. She concludes that despite numerous practical challenges, democratic approaches are crucial to our practice:

> Democratically oriented approaches to evaluation offer considerable promise. They are anchored in a profound acceptance of the intertwinement of values with facts in evaluative knowledge claims and the concomitant understanding that all evaluation is *interested* evaluation, serving some interests but not others. (Greene, 2006a, p. 136)

Greene (2002) draws on the work of Harry Boyte to conceptualize evaluation as a public craft that serves the common good by forging "a partnership between science and citizenship" (p. 9). She calls on evaluators to foster inclusive discourse that welcomes practitioner knowledge, privileges the lived experiences of program staff and participants, and engages directly and deeply with difference (Greene, 2002, 2005). While Greene (2005) values evaluators' technical expertise, she emphasizes that evaluation influences and is influenced by political decision-making and, therefore, cannot position itself as separate from political processes:

> Evaluation is an intrinsic part of the institutional fabric of public discourse and decision-making about public issues. We are among the weavers of this fabric, not just observers or admirers of the weaving. (p. 11)

As a public craft, Greene envisions evaluation that is fully engaged with urgent contemporary issues; "listens well" to foster inclusion, understanding, and appreciation of diverse viewpoints (Greene, 2005, p. 15); and legitimizes the "authentic knowledge" of program participants and communities (Greene et al., 2004, p. 101).

We would be remiss without also acknowledging the influence of Lee Cronbach's work on the "educative" strand to Greene's work:

> Cronbach's vision of evaluation has more to do with positioning evaluation to importantly inform broad, extensive discourse about enduring social prob-

lems and, thereby to advance societal betterment, than with championing a particular methodology or even methodological rigor per se as evaluation's most prized quality. (Greene, 2013a, p. 98)

Greene et al. (2004) aim to "stretch this tradition toward democratic ideals" (p. 99), emphasizing inclusion and engagement with difference, arguing that evaluation can serve democracy by surfacing and legitimizing multiple perspectives and realities.

ON PURPOSE, PUZZLES, AND THE PUBLIC GOOD IN GREENE'S WORK

When we reflect on Greene's democratic vision for evaluative and mixed-methods inquiry and our experiences of how she enacted this vision in day-to-day practices, three principles stand out. First, Greene challenges the tendencies of evaluators to use the purposes set by commissioners or funders to shape evaluations. Instead, she urges evaluators and researchers to craft clear and meaningful purposes to guide our work. Second, in the context of mixed-methods inquiry, she calls into question the privileging of triangulation and any assumed singularity to "truth." Instead, Greene encourages us to welcome empirical puzzles, look for the divergent and unexpected, and embrace inquiry as an iterative and creative process seeking pluralistic "truths." Third, she rejects the possibility of a value-neutral stance for researchers and evaluators. Greene views social inquiry as part of deliberative democracy with a professional responsibility to serve the public good. In the next section, we illustrate these three ideals.

Democratic Inquiry Guided by Purposes

Greene challenges the presumption that rationales of those commissioning the study, funders, or policy makers are (and should) motivate and frame evaluation and research. She also rejects the idea that such purposes are left implicit or unexamined by researchers. In evaluation and mixed methods, Greene pushes researchers and evaluators to commit to the rigorous work of crafting and naming purposes. This is in itself a democratic practice.

Evaluation

Greene and colleagues (2007) remind us that methodological decisions should follow from the study purposes and questions, which should follow from thoughtful consideration of multiple stakeholder and public interests:

There are multiple legitimate possible evaluation questions in every context... In an ideal democracy of engaged citizens, diverse stakeholders not only have the right but also the responsibility to participate in matters of governance and government policy. In real-world practice, a primary challenge of negotiations within the discretionary space of evaluation is to delineate a set of evaluation questions that honors the interests of evaluation sponsors but also makes space for the interests of other critical stakeholders. (p. 112)

In telling her own story of influences on her work, Greene (2015) recalls early professional experiences at the Curriculum Research and Development Center at the University of Rhode Island. It was there that her "disquiet" grew, later influencing her commitment to careful consideration of and explicit statement of the purposes that drive our work:

The main source of my disquiet was that the required evaluative information—while serving remote state and federal requirements—was not very relevant to the program directors and teachers who staffed these programs... These educators were most interested in how well the program was designed and implemented, in how various students experienced the program, and in how well the program supported diverse learning needs. (p. 747)

Greene grounds much of her evaluation work to prioritize the perspectives, questions, and information needs of the stakeholders who work to deliver and continuously improve the quality of their programs and address inequities. For example, in their values-engaged, educative evaluation approach, Greene et al. (2006) boldly call upon evaluation to name the value bases informing evaluation processes and products (see Boyce and Rivera, Section 1, Chapter 1).

Mixed Methods

Greene demonstrates how purposes ground and guide social inquiry within the mixed-methods field (Greene, 2006b). This begins with considering the broader sets of purposes that inform social and educational research and, specifically, those that we, as researchers and evaluators, choose to advance in our work.

All social science is framed and guided by inquirer assumptions about the nature of the social world, about what counts as warranted knowledge, about defensible methodology, and about the role of social inquiry in society, among others. It is simply not possible to conduct social inquiry without a self-understanding of the purpose and character of this activity. (Greene, 2013b, p. 111)

Greene's article with Valerie Caracelli and Wendy Graham (1989) sets forth core mixed-methods purposes to include triangulation, complementarity, development, initiation, and expansion. In her seminal text *Mixed Methods in Social Inquiry,* Greene (2007) later suggests that purposes are foundational to a "mixed methods way of thinking" (p. 20). She summarizes,

> To mix methods in social inquiry is to set a large table, to invite diverse ways of thinking and valuing to have a seat at the table, and to dialogue across differences respectfully and generatively toward deeper and enhanced understanding. (Greene, 2007, p. 14)

Democratic Inquiry Into Empirical Puzzles

Greene's (2007, 2013b) focus on democracy is also evident in her characterization of mixed methods as an opportunity to welcome empirical, controversial puzzles. Scholars have often positioned mixing of methods as a means of triangulation to assess whether data converged on similar findings or confirmed findings from one method or analysis to another (Fielding, 2012). Greene led the way interrogating early valuing of mixed methods for triangulation or confirmation. She continuously highlights research as interpretative and relational processes of inquiring and learning together. Greene's (2007) positioning of mixed methods as puzzles expands Tom Cook's (1985) conceptualization of critical multiplism—multiplicity of methods, perspectives, and analyses, among others, to "raise consciousness about what should be learned to help increase the likelihood that knowledge claims are true" (p. 46). Cook (1985) suggests "empirical puzzles" can be addressed by seeking to reconcile differences through examining and re-examining for points of agreement and convergence. Departing from this view, Greene reframes "empirical puzzles" as necessary for raising profound, and perhaps controversial, questions:

> Done thoughtfully, mixed-methods inquiry can generate puzzles and paradoxes, clashes and conflicts that, when pursued, can engender new perspectives and understandings, insights not previously imagined, knowledge with originality and artistry. Surely the continuing crises that plague our world need all of our creativity and insight. (Greene, 2007, p. 24)

Greene thus moves forward our thinking, imploring scholars to critically consider why and how we approach the empirical puzzles that we speak to in our scholarship.

Democratic Inquiry in Service to the Public Good

Departing from the post-positivist, values-free paradigm prevalent in social research and evaluation (House & Howe, 1999; Shadish et al., 1991),

Greene's research asserts that evaluation is fundamentally intertwined with values and inevitably advances particular values and interests (Greene, 1997; Greene & McClintock, 1991). Accordingly, she emphasizes the public good—and advocacy for marginalized peoples and societies—as a central goal of research and evaluation (Greene, 2013b). In doing so, Greene's scholarship boldly and persistently calls on researchers and evaluators working across substantive areas, methodological traditions, and institutional settings to remember and take responsibility to ensure that social inquiry serves the public good.

Her paper aptly titled "Evaluation as Advocacy" (Greene, 1997) warns against assumptions or claims of value-neutrality in which researchers can opt out or remain impartial to the interests, values, and normative ends of research. Instead, she argues for an interpretation of "advocacy" not as an impassioned, unquestioning allegiance to some stakeholder groups, but as an intentional claiming of the sociopolitical, ideological, and value commitments shaping one's work. Overall, she encourages researchers to "claim a louder voice in contemporary conversations about social problems and policies" to garner "a more active claim to scientific citizenship, envisioned herein as the assumption of public accountability and social responsibility for the political, moral, and value consequences of one's work as scientists" (Greene, 2015, p. 748).

DEMOCRATIC IDEALS IN GREENE'S TEACHING

"How do you want to tell the story? Remember, it may be your findings, but it is their story." This signature Greene "question-reminder" grew familiar to us during our doctoral studies at the University of Illinois at Urbana-Champaign (Illinois; where we studied with Williams and Mustafaa, Section 1/Chapter 2). As doctoral students in educational psychology (Gates, Teasdale, and Hidalgo Standen) and educational administration and leadership (La Londe), we took two of her courses, Mixed-Methods Inquiry in Social Research and Evaluation (mixed methods; Gates, La Londe, and Hidalgo Standen in 2013; Teasdale in 2015) and Evaluation Methods (Gates, La Londe, and Hidalgo Standen; Teasdale in 2015). In Greene's 2013 Evaluation Methods course, students conducted evaluations in small groups of different university and community initiatives. Greene partnered with faculty in the School of Engineering, and students evaluated the school's service-learning courses for undergraduates Learning in Community (LINC). LINC students engaged in interdisciplinary research, service, and fieldwork activities with local and international community partners. Over approximately 3 months, Greene and teaching assistant Dr. Nora Gannon-Slater, led us through research design, instruments, data collection and analysis, and communication of findings to the LINC directors and other evaluation stakeholders. We practiced the nuts and bolts of evaluation methods and critically reflected on the ideals and rationales guiding our work.

Greene's mixed-methods course centered on challenging questions: What exactly are we mixing, and for what purpose? What will we do when the mixing gets messy and when the mixing leads to puzzling contradictions? How will we approach findings that are unfavorable or hard to digest regarding major social issues, such as poverty and housing inequity? These moved us away from "how-to-do mixed methods" toward a provocative examination of the possibilities of democratic inquiry in mixed-method research. With Greene's (2007) *Mixed Methods in Social Inquiry* as an anchor text, we participated in discussions and exercises that engaged theoretical, substantive, and methodological issues in democratic mixed-methods inquiry. Two discussions particularly stand out in our memories. We discussed an issue that Trend (1979) raised when participant observation data could not be reconciled with quantitative, administrative data. The article challenges the idea that multiple methods generate data that can easily be combined and instead suggests that "the neat dovetailing of a research puzzle should be cause for suspicion" (p. 345). Our examination of the Kling et al. (2005) article, "Bullets Don't Got No Name: Consequences of Fear in the Ghetto," raised issues surrounding observations and in-depth interviews in experimental research on housing assistance for low-income families. Discussing these studies helped us see mixing methods as a worthwhile opportunity for democratizing inquiry.

OUR REFLECTIONS ON JENNIFER C. GREENE

As we have discussed, democratic ideals are woven throughout Greene's scholarship and teaching. In this section, we offer reflections on how Greene's democratic ideals shape our scholarship and teaching.

Reflection by Gates

In 2011, I moved to Illinois to pursue a PhD with Jennifer as my advisor. In searching for a graduate program, I spoke with faculty and students at various programs, sharing my interests and listening to their advice. It was in those conversations that I learned of Jennifer Greene and her mixed-methods scholarship and the program at Illinois as a pillar of democratic and responsive evaluation—and what a place it was.

What stands out most about my time with Jennifer is how she practiced her ideals in such a down-to-earth way. For someone who has significant and influential ideas about evaluation and mixed methods, she positioned herself as one voice in a dialogue. Being in conversation with her meant you could count on her to listen well, ask probing questions, and uplift your thinking and spirit. She genuinely welcomed interdisciplinarity and pursuit of new directions. Whereas some faculty want students to carry on *their* research, Jennifer supported her students in pursuing what they deeply cared about. She also pushed us to consider how we could contribute to theoretical, methodological, or practical challenges in the field. In my case, this meant trying out different interests until I found systems thinking, which struck an immediate spark. She encouraged me to pursue training, talk with experts, and even try out systems approaches in our evaluation projects. And she did this for each of her advisees—encouraging them to bring their commitments (see Davis & Kaminsky, Section 1/Chapter 4) and pursue the purposes motivating their work.

It was really when I began teaching and mentoring students that I realized the extent to which Jennifer shaped my sense of what it means to be a scholar-practitioner (see Schwandt, Section 3/Chapter 6) and mentor. As an assistant professor in the Measurement, Evaluation, Statistics, and Assessment Department at Boston College, I teach mixed methods using Jennifer Jennifer's book, *Mixed Methods in Social Inquiry* (Greene, 2007) drawing on my experiences with her course. As students begin their coursework eager to learn *how* to do research and evaluation, I get the honor and challenge of holding spaces for questions about the purpose(s) evaluations should serve and exposing them to differing perspectives within the field. Challenging research ideas simply or solely as a pursuit of truth, I turn to her work to convey roles for evaluation in learning and dialoging about critical social issues, what we can and should do about them, and how this can mean holding multiple and sometimes contradictory truths and perspectives.

Reflection by La Londe

I am beyond fortunate and grateful for Jennifer's mentorship and teaching. Apart from the abovementioned courses, I grew tremendously from Jennifer's mentorship when she was a member of my doctoral committee, as well as when I was a research assistant on her Spencer Foundation-funded project, "The Role of Student Characteristics in Teachers' Formative Interpretation and Use of Student Performance Data" (Gannon-Slater et al., 2017)

As a member of Jennifer's research team, I gained invaluable principles about habits of mind, development of scholarship, and teaching and mentorship principles that I have applied to my career. Jennifer embraced consistency, clarity, and community. We met every Thursday at 8:30 a.m., every meeting had a clear agenda, and every meeting started with informal check-ins. Jennifer facilitated a collaborative, focused ethos, where our expertise and stances became the starting points for analyses. When challenging, courageous conversations were necessary, Jennifer embodied care and a commitment to diversity in voices.

I have also carried with me Jennifer's invaluable guidance on scholarship. Her feedback on our collective writing in the research project and on my dissertation sedimented core principles that guide my scholarship and mentorship to doctoral students. As a result, I focus on framing, clarity, and tailoring stories to audiences. Jennifer's mentorship has also guided my navigation of the tenure track as a mother of young children. I often remind myself of Jennifer's wisdom, "Don't add a layer of angst. It's not helpful." Her casual comments in many of our conversations led me to a place where today I remind myself and my scholar mama friends to "remember the bigger picture" and find ways to make the "mix" of parenting and career work by finding a common purpose.

Perhaps the lesson that I am most grateful for from Jennifer is the focus of this chapter—purposes, puzzles, and the public good. Throughout my doctoral studies and in my conversations with Jennifer in my early career, I learned the value of thinking about *how* and *why* I make the research design decisions I make and remaining aware of the merits and consequences of the purposes I deem important. She helped me come to a place where I know my purpose and know *how* to think about complex puzzles. I know my small dent in teaching and research that serves the public good is indeed just a small dent, but this is necessary and must not be underestimated.

Reflection by Teasdale

I met Jennifer in 2013 when I was investigating doctoral programs, and I came to Illinois in 2014 to pursue my PhD under her mentorship. Beginning with our first conversation and stretching across the 5 years I studied with Jennifer, she taught me that research and evaluation methods shape knowledge and that values drive methods. She showed me how values are expressed in choices about study purposes and through subsequent decisions about evaluation questions, other study components, and our interactions with stakeholders.

My understanding of how purposes, values, and methods are intertwined unfolded through work as one of Jennifer's research assistants—as she named the values that underpinned our study of teachers' use of student performance data and selected methods that advanced those values—and in her evaluation and mixed-methods classes, in which she guided her students in identifying the values implicit in the studies we critiqued. In

2016, I had the privilege of exploring these ideas further in her advanced doctoral seminar focused on values in evaluation practice. Through our readings and discussion, Jennifer highlighted how values are embedded (and hidden) in evaluation methodologies and how, by addressing specific purposes, evaluation inevitably advances particular interests. She led our team of PhD students in an empirical study of evaluators' values, teaching us to surface, identify, and investigate values through inductive, reflective inquiry that examined individuals' beliefs and behaviors within the context of their lives and experiences. Across these classes, projects, and conversations, Jennifer made visible the specific ways that values shape the purpose of inquiry and challenged us to advance democratic values and, ultimately, the public good in our own work.

As Jennifer's student, I was afforded the opportunity to deeply explore the intersection of values, methods, and democratic visions for evaluation—and wrestle with the implications for serving the public good. Jennifer encouraged me to forge my own path through this territory, defining a purpose for my research agenda and approach to evaluation practice that is uniquely my own, while firmly grounded in the training and mentoring she gave me. Now, as an assistant professor of evaluation, I endeavor to carry on her legacy by attending to the centrality of values in evaluative inquiry and advancing democratic aims. I guide my students in examining the intertwining of values and methods, recognizing how all evaluations inevitably advance particular interests, and clarifying and articulating the values they hold. Guided by Jennifer's reminder that there are multiple possible evaluation questions in every context (Greene et al., 2007), my research group investigates evaluative criteria to understand which purposes and values—and whose purposes and values—those questions represent.

Perhaps even more importantly, I try to carry on Jennifer's legacy by striving to emulate her day-to-day interaction with students and colleagues. As her student, I saw Jennifer consistently demonstrate kindness, patience, and caring, encourage depth of thought about important issues, and hold space for diverse perspectives and opinions. Speaking about evaluative findings, Jennifer (2005) explains: "What we do and how we interact in a setting matters to what we learn. These relationships become constitutive of our evaluation findings" (p. 14). She demonstrated the central importance of relationships in her teaching, mentoring, and research, as well. In doing so, Jennifer serves as a role model for the kind of professor I aim to be.

Reflection by Hidalgo Standen

In 2009, I obtained a scholarship from the government of Chile and the Fulbright Foundation to pursue doctoral studies. I lived for 4 years in

Illinois and had the honor of studying with Jennifer, my mentor, teacher, and director of my dissertation, "Teaching Quality in La Araucanía Chile" (Hidalgo Standen, 2020). During those years, I had the opportunity to study in-depth the theories and methods that support the evaluative discipline, and I could learn about the founding work of Jennifer in the area of mixed methods. One of the most important things I learned during my doctoral program was from Jennifer's integrity and generosity; she lives and works so consistently with the principles that she writes about, teaches, and practices in her evaluation work.

Looking back, I realize three ideas have marked my work in a particular way. Regarding purpose, evaluation should advance the values of a democratic society. This idea implies understanding evaluation not only as a technical process but also as a political process in itself. Evaluation as a discipline should engage with political processes of building a more just and equitable society. This guiding principle profoundly influences those who work as evaluators, since it implies rethinking methodological decisions during different stages of the evaluation process to ensure that the evaluation favors the active participation of the community, giving space to multiple voices. This option is not without difficulties, but I try to actively respect different points of view and search for common languages that facilitate communication and mutual understanding.

A second idea, within the framework of the study of mixed methods, is Jennifer's idea of mental models (Greene, 2007). Specifically, she encourages researchers to explain their "mental maps" understood as the theoretical, political, ideological influences that underpin their research decisions. Currently, I am working on a research project to analyze different ways of learning from Mapuche indigenous children in southern Chile. The assumption that guides this research is that the indigenous heritage and culture of the Mapuche children shape their way of understanding the world and their ways of interaction in situations of formal and informal learning. In this project, I work with a multidisciplinary team composed of psychologists, an anthropologist, a teacher educator, a Mapuche researcher, and me. My training has allowed me to understand that the diversity of opinions and points of view is an opportunity to strengthen collective work, generate synergy between teams, look for meeting points, and value disagreements as learning opportunities.

Finally, I want to highlight what I think is one of Jennifer's main contributions. The progress in the construction of a democratic society is not only dependent on the epistemological and political orientation of our work, but also a commitment that must be expressed in the day-to-day activities of our actions and our interactions with others.

FORWARDING JENNIFER GREENE'S LEGACY

Jennifer Greene's name and scholarship are widely known and highly regarded in the evaluation and mixed-methods fields. We surmise Greene would want her mentees to invite our current and future students to approach her scholarship with care and an attitude of critical reflection on their own values, stances, and questions. We thus invite rising and future scholars to understand Jennifer C. Greene as one who envisions and enacts evaluation and mixed-methods research as practices driven by clear purposes, welcoming of puzzles, and in service to the public good. She grounds these commitments in democratic traditions (Greene, 2006a), valuing difference and dialogue as means for "society to learn about itself" (Cronbach, 1980, p. 2) and just what it means to be a "good" researcher.

In our research, teaching, and mentorship, we are constantly reminded and challenged by Greene's main encouragements:

• Be thoughtful and explicit in claiming the *purposes* served in your work. Make sure the interests, values, and potential consequences are in line with your commitments.
• Remember that inquiry into complex phenomena inevitably raises *puzzles.* Do not get caught up proving, confirming, and agreeing; rather, welcome uncertainty, raise new questions, and seek out differences.
• Ultimately, we serve the *public good.* This means we must look beyond the immediate priorities and scope of our work, often the one shaped by funding agencies and policy makers, to consider the diversity of wider stakeholder and public interests and, when possible, incorporate participation and deliberation into our work.

As we write this chapter amidst the volatility of today's anti-Black, anti-people of color, and COVID-19 pandemic, we are acutely aware that Greene's encouragements are vital to our contributions to social and educational research.

We close with a tribute to friendship as, perhaps, her greatest legacy. She brought the four of us together through her courses, research projects, the Illinois evaluation program, and a much wider network of lifelong friends. In the words of Greene (2015),

> Many of my best friends and valued evaluation colleagues persisted in methodological allegiances that were other than my own. We must never underestimate the importance of relationships in all life's domains, even a domain as esoteric and obtuse as methodological commitments. (p. 748)

We take the friendships made through our connections with Greene to be of the most vital, and so we close with one last guidance: Let's center our work in the relationships we build, taking care not to treat these relationships as means (for access, data, use, etc.) but as ends in themselves (Greene, 2001).

REFERENCES

Cook, T. D. (1985). Post-positivist critical multiplism. In W. R. Shadish & C. S. Reichardt (Eds.), *Reproduced in evaluation studies review annual* (Vol. 12). SAGE Publications.

Cronbach, L. J., & Associates. (1980). *Toward reform of program evaluation.* Jossey-Bass.

Fielding, N. G. (2012). Triangulation and mixed-methods designs: Data integration with new research technologies. *Journal of Mixed Methods Research, 6*(2), 124–136. https://doi.org/10.1177/1558689812437101

Gannon-Slater, N., La Londe, P. G., Crenshaw, H. L., Evans, M. E., Greene, J. C., & Schwandt, T. A. (2017). Advancing equity in accountability and organizational cultures of data use. *Journal of Educational Administration, 55*(4), 361–375. https://doi.org/10.1108/JEA-09-2016-0108

Greene, J. C. (1997). Evaluation as advocacy. *Evaluation Practice, 18*(1), 25–35. https://doi.org/10.1177/109821409701800103

Greene, J. C. (2001). Dialogue in evaluation: A relational perspective. *Evaluation, 7*(2), 181–187. https://doi.org/10.1177/135638900100700203

Greene, J. C. (2002). *Towards evaluation as a "public craft" and evaluators as stewards of the public good or on listening well* [Paper presentation]. 2002 Australasian Evaluation Society International Conference, Wollongong, Australia.

Greene, J. C. (2005). Evaluators as stewards of the public good. In S. Hood, R. K. Hopson, & H. Frierson (Eds.), *The role of culture and cultural context: A mandate for inclusion, the discovery of truth, and understanding in evaluative theory and practice* (pp. 7–20). Information Age Publishing.

Greene, J. C. (2006a). Evaluation, democracy, and social change. In I. F. Shaw, J. C. Greene, & M. M. Mark (Eds.), *The SAGE Handbook of Evaluation* (pp. 119–140). SAGE Publications.

Greene, J. C. (2006b). Toward a methodology of mixed methods social inquiry. *Research in the Schools, 13*(1), 93–99.

Greene, J. C. (2007). *Mixed methods in social inquiry.* Jossey-Bass.

Greene, J. C. (2013a). The educative evaluator: An interpretation of Lee J. Cronbach's vision of evaluation. In M. C. Alkin (Ed.), *Evaluation roots: A wider perspective of theorists' views and influences* (pp. 170–179). SAGE Publications.

Greene, J. C. (2013b). Reflections and ruminations. *New Directions for Evaluation, 2013*(138), 109–119. https://doi.org/10.1002/ev.20062

Greene, J. C. (2015). The emergence of mixing methods in the field of evaluation. *Qualitative Health Research, 25*(6), 746–750. https://doi.org/10.1177/1049732315576499

Greene, J. C., Caracelli, V. J., & Graham, W. F. (1989). Toward a conceptual framework for mixed-method evaluation designs. *Educational Evaluation and Policy Analysis, 11*(3), 255–274. https://doi.org/10.3102/01623737011003255

Greene, J. C., DeStefano, L., Burgon, H., & Hall, J. (2006). An educative, values-engaged approach to evaluating STEM educational programs. *New Directions for Evaluation, 2006*(109), 53–71. https://doi.org/10.1002/ev.178

Greene, J. C., Lipsey, M. W., Schwandt, T. A., Smith, N. L., & Tharp, R. G. (2007). Method choice: Five discussant commentaries. *New Directions for Evaluation, 2007*(113), 111–127. https://doi.org/10.1002/ev.218

Greene, J. C., & McClintock, C. (1991). The evolution of evaluation methodology. *Theory Into Practice, 30*(1), 13–21. https://doi.org/10.1080/00405849109543471

Greene, J. C., Millett, R. A., & Hopson, R. K. (2004). Evaluation as a democratizing practice. In M. T. Braverman, N. A. Constantine, & J. K. Slater (Eds.), *Foundations and evaluation: Contexts and practices for effective philanthropy* (pp. 96–118). Jossey-Bass.

Hidalgo Standen, C. (2021). The use of photo elicitation for understanding the complexity of teaching: A methodological contribution. *International Journal of Research & Method in Education, 44*(5), 506–518. https://doi.org/10.1080/1743727X.2021.1881056

House, E. R., & Howe, K. R. (1999). *Values in evaluation and social research.* Sage.

Kling, J. R., Liebman J. B., & Katz L. F. (2005). Bullets don't got no name: Consequences of fear in the ghetto. In T. S. Weisner (Ed.), *Discovering successful pathways in children's development: New methods in the study of childhood and family life* (pp. 243–282). University of Chicago Press.

Shadish, W. R, Cook, T. D., & Leviton, L. C. (1991). *Foundations of program evaluation: Theories of practice.* Sage.

Trend, M. G. (1979). On the reconciliation of qualitative and quantitative analyses: A case study. In T. D. Cook & C. S. Reichardt (Eds.), *Qualitative and quantitative methods in evaluation research* (pp. 68–86). SAGE Publications.

CHAPTER 6

VISITING AND LISTENING WELL

A Tribute to Jennifer C. Greene, the Gentle Disruptor

Veronica G. Thomas
Howard University

ABSTRACT

This chapter examines Greene's stance and concrete ways she democratizes evaluation practice by emphasizing inclusion, collaboration, cultural responsiveness, and educative inquiry while respectfully striving to achieve genuine and active stakeholder engagement. I begin the chapter by situating Greene's stance within the broader context of her upbringing and her coming into adolescence and young adulthood during the height of major social movements of that period. These important events undoubtedly influenced her evaluative thinking and calls for a more socially just, inclusive, and culturally responsive evaluation (CRE) practice. Next, I share my initial and most memorable encounters with Greene to exemplify her as a "disruptor," someone who did things differently and continues to champion increasing diversity within the field. Subsequently, in discussing Greene's scholarship and practice, the over-

Disrupting Program Evaluation and Mixed Methods Research for a More Just Society, pages 77–93
Copyright © 2023 by Information Age Publishing
www.infoagepub.com

arching theme of this chapter is her notion of "visiting and listening well." The premise here is that evaluators must be mindful that they are visitors in others' spaces and, as such, they should behave with respect, honesty, and trustworthiness. The chapter concludes with a summary of how Greene re-envisioned the role of evaluation and the evaluator, and how she continues to be a beacon for promoting diversity, equity, and engagement in evaluation and within the broader society.{/ABS}

Throughout her professional career including her teachings, speeches, writings, and practice, Jennifer C. Greene has been a "gentle disruptor." She challenged the status quo by offering new perspectives and being unapologetic in surfacing values in her work.

Greene, evaluation scholar practitioner, has long argued that evaluators must "visit and listen well" and that evaluations should privilege values of equity, justice, inclusion, and voice, as these democratic values are, in her words, "most defensible." Further, she contends that the evaluator has both the opportunity and the responsibility to advance these core democratic values as a primary basis for evaluative actions and judgments. The evaluator's job, according to Greene, is to assertively, and always politely and respectfully, legitimize the core democratic evaluation principle that all stakeholders (or their designated representatives) should have an opportunity to inform, shape, and interpret the process and product of the evaluation. It was Greene who vividly proclaimed that this responsibility "requires action, and not just talk" ("Voices from the Field Interview" published in Thomas & Campbell, 2021, p. 254).

This chapter examines Greene's stance and concrete ways she democratizes evaluation practice by emphasizing inclusion, collaboration, cultural responsiveness, and educative inquiry while striving to achieve genuine and active stakeholders' engagement in respectful ways. I begin the chapter by situating Greene's stance within the broader context of her upbringing and her coming into adolescence and young adulthood during the height of major social movements of that time. These important events undoubtedly influenced her evaluative thinking and calls for a more socially just, inclusive, and culturally responsive evaluation (CRE) practice. Next, I share my initial and most memorable encounters with Greene to exemplify her as a "disruptor," someone who did things differently and continues to champion increasing diversity within the field. Subsequently, in discussing Greene's scholarship and practice, the overarching theme of this chapter is her notion of "visiting and listening well." The premise here is that evaluators must be mindful that they are visitors in others' spaces and, as such, they should behave with respect, honesty, and trustworthiness. The chapter concludes with a summary of how Greene re-envisioned the role of evaluation and the evaluator and how she continues to be a beacon for promoting diversity, equity, and engagement in evaluation and within the broader society.

ON BEING JENNIFER C. GREENE: INTEGRATING THE PERSONAL AND THE PROFESSIONAL

In an article, "In Gratitude to Jim and Whit," published in *New Directions for Evaluation*, Greene (2018a) recounts how she learned compassion for others and the importance of public education by watching her parents and participating in student protests and marches in Washington, DC, during the 1960s. She noted that her parents did not tell her about compassion or preach about it; they just did it. Instead, Greene (2018a) learned about compassion simply by observing what her parents did in their everyday lives and how they acted as "good citizens" (p. 75) and "good community people" (p. 75). She recalls those early times being followed by pivotal years during the 1960s that she spent, in part, participating in student protests advocating for a more just and equitable American society.

During the height of the civil rights, women's, peace, and student movements of the 1960s and 1970s, many young people of various ethnicities and races became activists for social justice. Supporters of these movements questioned why and how different groups were treated (or mistreated) and rallied to address many inequities. Many young people, such as Greene, participated in these movements not only to shape their own future but to cultivate a more socially, politically, and economically just country for those most often discriminated against and shut off from opportunities and access to all the benefits and privileges of being an American citizen.

In keynote addresses, presentations, and panel discussions, Greene continues to recount her experiences growing up during the revolutionary times of the 1960s with "liberal, progressive sensibilities" (University of Illinois Urbana-Champaign, 2017, para. 8). Speaking on a 2017 student-organized panel about diversity in STEM, sponsored by the University of Illinois–Champaign Science Policy Group (a registered student organization), Greene described her involvement in protest during those times as follows: "One week was a yellow arm band; the next week was a red one; and the next one a green one; it was all very exciting" (University of Illinois Urbana-Champaign, 2017, para. 7). Her involvement in protests, she notes, had a significant impact on her belief system, stating, "You develop some sensibilities about what's right and what's wrong with the direction of the country when you're in the midst of something like that" (University of Illinois Urbana-Champaign, 2017, para. 8) . Greene points out that those beliefs also helped shape her career choice as an educational psychologist/researcher who developed a keen interest in how our educational system could better serve everyone, particularly those least served. Greene gained useful ideas and practical skills as a "young activist" that subsequently impacted her thinking and practice as an evaluator.

Early in her career as an evaluator, Greene did not see evaluation as a political activity. However, over the years, she indicates that she conceptualized the integration and connection of political beliefs and values within her evaluation scholarship and practice. Today, Greene is adamant in her proclamation that evaluation is not neutral but, instead, political since it advances some values and not others. In a 2010 interview, Greene pointed out that evaluators are free to pick the values to which they want to commit; she, however, cautioned them that opting in favor of one value or another should be based on the role they wish to play in society rather than on a partisan stance toward a particular program (Tarsilla, 2010). As Greene continued to grow in the evaluation field, she discovered that she could infuse value commitments toward public good, equal rights, and equity into her evaluation practices which, as she argues, are conducted on behalf of others, especially those on society's margins. For Greene (2005a), evaluation in service of the public good entails a technically proficient and politically and socially responsible practice and one where evaluators play a social enfranchising role.

My Initial Encounters With Greene

Greene acknowledges she has lived a life of incredible privilege compared to the majority of people in the world. To her credit, she has been relentless in using whatever privileges she has to give back, contribute, and make better. For Greene, this is not just talk; it is backed up with action aimed at democratizing evaluative inquiry toward greater inclusiveness, collaboration, and learning. Below, I describe my initial personal encounters with Greene that support this assertion.

I first met Greene in 2002 during an American Evaluation Association Annual Conference when she was co-editor of *New Directions for Evaluation* (NDE). Sometime the following year, I submitted a proposal for an *NDE*-focused issue on co-constructing a contextually responsive evaluation framework. I wrote the proposal with my colleagues at the (then) Howard University Center for Research on the Education of Students Placed at Risk (CRESPAR). Greene vigorously mentored me through this process, from beginning to end. On May 21, 2003, I had a 2-hour conference call with Greene (and incoming *NDE* editor Jean King) about the volume and the various contributing chapters. Subsequently, on December 2, 2003, Greene sent me a detailed letter, slightly over five pages in length, with editorial feedback on what was specifically needed to address the reviewers' comments as well as her own comments as editor to elevate the manuscript to publication quality.

Never had I encountered a journal editor, particularly someone of Greene's professional stature, to provide such detailed feedback (verbally and in writing). Never had I encountered a journal editor so generous with her time, so dedicated to the field, and so committed to elevating the work of others in the discipline, particularly those often not heard. Needless to say, I was star-struck! To me, this is part of Greene's disruptive nature, doing things differently than what was usual to expand equity perspectives, add new and diverse voices and scholars in the field, and foster collaboration among unlikely collaborators.

A few noteworthy comments in that December 2, 2003, communication from Greene that continue to shape my evaluative thinking today and shine a spotlight on Greene's perspective include:

> It seems that your evaluation approach is clearly on the side of fuzzier boundaries. In fact, a major and distinctive strength of your evaluation approach is perhaps the notion of one integrated program/evaluation team, who are collectively responsible and accountable for program development, implementation, and evaluation. This notion suggests that the functions of program intervention in urban schools serving children of color are pretty much the same as the functions of evaluation and so a conflation of these activities makes sense. This notion positions evaluation and evaluators as partly responsible for program impacts and outcomes.
>
> And this is a pretty radical idea that may be especially important in traditionally underserved communities. Evaluators are not neutral actors in these contexts, but either support or don't support, through their activities, the intentions of the program they are evaluating. Given the politics of difference in U.S. society today, one has to take a stance. (personal communication, December 2, 2003)

Greene embraced how contributing authors proclaimed their interconnectedness with program staff, their shared lived experience with people in the community under study, and their aim toward doing evaluations "with" and not simply in the community. The result of our diligent work, coupled with Greene's commitment toward expanding perspectives and diverse voices in the evaluation field, resulted in the first issue of *NDE*, edited by two African American women, Floraline I. Stevens and me, with all the volume's contributing authors being African Americans scholars and practitioners who were passionate about planning and conducting evaluations with five overlapping themes: engaging stakeholders, co-construction, cultural and contextual relevance, responsiveness, and triangulation of conceptual and methodological perspectives (Thomas & Stevens, 2004). These five themes are consistent with Greene's evaluation stance. Her approach heavily influenced many of the ideas espoused in this issue of *NDE*, as evident by the numerous times her work is cited throughout the various chapters.

JENNIFER C. GREENE ON MAKING WISE DECISIONS

From my own perspective, wise decision-making, in general, and in the case of evaluation, involves the evaluator's (and funder's) willingness toward looking for and seeing the bigger picture. This can be best achieved, in large part, by taking a broader perspective that allows one to consider multiple aspects (e.g., historical, sociocultural, economic, political) of a situation from different angles and from the perspective of the different people and not only the most powerful or vocal. Greene's emphasis on "wise decision-making" aligns with this perspective. She argues that wise decision-making is informed by including more stakeholder perspectives, interests, concerns, and values than are typically included. This means respectfully considering the concerns of all represented in the project and paying particular attention to diverse points of view not necessarily reflecting the dominant paradigm or the perspectives of those in positions of power and authority.

Greene views good evaluation as making wise decisions or judgments with many dimensions most appropriate for the context at hand. Understanding of context, as she asserts, is essential to wise technical decision-making in selecting, adapting, and creating measures to fit the norms, language, and values of the context of the evaluand. Greene (2018b) says it is important for an evaluator to "know when to insist upon using field-tested interview guides or psychometrically-sound questionnaires, and when the characteristics of a particular context are more important than remote indicators of measurement quality" (para. 1).

Wise decision-making in evaluation can be facilitated through conversations and negotiations between the evaluator and (often) diverse stakeholders (Greene, 2018b). As such, attending to the social and interpersonal dimensions of evaluation is central, in Greene's perspective, to enabling wise decisions and taking wise actions. This certainly was not a perspective that was generally accepted in the evaluation field several decades ago.

EVALUATION BEING A SOCIAL PRACTICE OF VALUING

In a CREA (Center for Culturally Responsive Evaluation and Assessment) blog, Greene (2016) emphasized that evaluation is the social practice of gathering and using empirical information to judge the merit and worth of a given social or educational intervention somewhere in the world. She adds that our empirical data and judgments are grounded in particular values that are too often "unnamed" and thereby underused to advance our culturally responsive agenda of cultural respect and sociopolitical equity. Greene spoke boldly of engaging values in evaluation when many in the field were still grappling with even acknowledging values in their work. In fact, she refers to her approach as values-engaged because it explicitly and intentionally involves both descriptions and prescriptions of values within evaluation (Greene, 2005b; Greene et al., 2006).

At the heart of much of Greene's scholarship is privileging, engaging, and advancing the democratic ideal of equity. Equity, for Greene, can be enacted through genuine concern with the treatment of program stakeholders and paying attention to how well the program affords program access to diverse participants, soliciting meaningful participation from diverse stakeholders through dialogue and other means, and considering the consequences of the program for all individuals and groups that are present in the context, particularly those least served (Hall et al., 2012). Her values-engaged approach has been applied in various projects, and there are many journal articles, book chapters, and even a guidebook for practice that describes this approach (e.g., Greene et al., 2011; Hall et al., 2012).

The Fallacy of Neutrality

For hundreds of years, philosophers of science have commented on the difficulty of attaining scientific objectivity. As early as the 1800s, Isaac Watts, writer and theologian, aptly said, "The eyes of a man in the jaundice make yellow observations on everything; and the soul tinctured with any passion diffuses a false color over the appearance of things" (AZ Quotes, n.d., para. 1). Nevertheless, many evaluators (and researchers), most often like to think of themselves as fairly objective and unbiased. Worthen (2004) argues that for most evaluation work to be useful, the evaluator needs to be "sufficiently neutral, nonpartisan, and dispassionate about that which is evaluated to avoid unrecognized biases from coloring his/her portrayals and evaluative judgments in ways that alter facts or dilute reality" (p. 414).

Objectivity and unbiasedness do not exist in scientific research or practice. Greene's (2002) response to this issue is that evaluators cannot practice within a protected air bubble in the hopes that their work will not perturb the situation or "influence it via some form of unwanted bias" (p. 2). We all see the world through our own lens, colored by our lives, experiences, hopes, and fears. Greene passionately argues that evaluation, like all forms of social inquiry, is "at least partly constitutive of the context in which it takes place, particularly of the institutional and interpersonal relationships of power, authority, and voice in that context" (p. 2). And, she has boldly pronounced that

> like it or not, the practice of evaluation itself either sanctions and reinforces, or alternatively challenges and disrupts, key dimensions of these contexts. Notably:
> - Who has the right to be heard about what?
> - What counts as legitimate knowledge?
> - How decisions are made—who participates, what happens publicly, and what happens behind the scenes?
> - What factors or criteria are valued in making decisions, and who gets to determine these?
> - And also, the ways that people relate in a given context—with trust or suspicion, respect or disregard, reciprocity or selfishness, caring or neglect. (Greene, 2002, p. 3)

While many evaluators were arguing for objectivity, or at least the appearance of objectivity, Greene, as a gentle disruptor, was part of a growing chorus of evaluators in the 1990s and early 2000s, who acknowledged the impracticality of objectivity as a stance in the field. In fact, she expresses that a stance of "distanced neutrality" means support for the current state of affairs. Greene, without reservation, asserts that she chooses to do evaluation with a conscious choice and a commitment to democratic values, making judgments of program quality that are contextually respectful and relevant, and foregrounding democratic concerns about equity and justice in her work. Her perspective has given many of today's evaluators the courage to reject the fallacy of objectivity, the voice to take a stance and disclose their own values and cultures, and the understanding of how their stance shows up in their work.

A Practice of Valuing

Greene's values-engaged approach to evaluation involves both describing stakeholder values, which is a common practice in responsive evaluation, and prescribing or advocating certain values with an emphasis on inclusion and equity. She asserts that values of diversity and equity can be prescriptively and dialogically advanced through multiple strategies. Two strategies include explicitly (a) acknowledging the evaluation's equity orientation and (b) raising concerns about equity in the particular context throughout the entire evaluation process, such as asking questions regarding the extent to which a program has been successful at providing opportunities and accomplishments for all participants, with particular emphasis on individuals from groups traditionally underserved. Greene also recommends defining the quality of programs at the intersection of program content, pedagogy, and equity. This should be coupled with attending to the interests, perspectives, and relationships among multiple and diverse stakeholders throughout the evaluation process as important ways to prescriptively advance values of diversity and equity. This perspective has influenced the field considerably as an increasing number of evaluators explicitly acknowledge how their values influence the choices they make, the conclusions they draw, and the interpretations they reach in an evaluation.

EQUITY AND INCLUSION THROUGH DEMOCRATIC APPROACHES AND CULTURALLY RESPONSIVE EVALUATION

At a time when so many others in the evaluation field were focusing primarily on objectivity and methodological issues, Greene took a different view. She intertwined methodology and values. She clearly respects evaluation as a methodological and technical enterprise. However, she argues that it is more fundamentally a social practice that, at its best, respectfully and explicitly engages with the diversity of viewpoints, experiences, stances, and values. For Greene, this responsive approach to evaluation is best realized when the evaluation is attuned to the issues and concerns of people within the context of the program under study. In a fairly recent interview that I conducted with Greene (personal communication, November 14, 2019), she stated,

> You can't assume you know much, and your wish is to know more as you engage stakeholders. It can be tricky at times. But try to be as completely honest and transparent as possible with stakeholders ensuring that there is no hidden agenda—that your words are the truth.

Democratic Approaches to Evaluation

Following the democratic ideas of Ernest House and colleagues (e.g., House, 2014; House & Howe, 1999), Greene exhorts evaluators to equitably include representatives from all important stakeholder groups, including decision-makers, administrators, program participants, and community members. She stresses that evaluation, in some respects, is an intervention itself that can disturb the status quo in ways necessary for society to reach its full democratic ideals (Greene, 2015). Although Greene acknowledges that many may not agree with her support for a democratic approach to evaluation, she encourages evaluators to recognize the sociopolitical intervention strands of their own work through the advancement of stakeholder interests and values, represented most evident in the evaluation's purpose, audience, and key questions, and quality criteria upon which judgments of program quality and effectiveness are made (Greene, 2015). She boldly argues that evaluators should name, claim, and justify the values that are advanced by their evaluation practice.

Throughout decades of evaluation scholarship and practice, Greene re-envisions evaluation practice in ways that many others in the field dared not tread. Her accomplishments are borne out of her beliefs, as she pronounced in her keynote address at the Australasian Evaluation Society International Conference:

> Evaluation cannot position itself on the sidelines of political decision making... [E]valuation cannot be a bystander of democratic discourse; it is inevitably a player therein. And so strong, worthy, powerful evaluation is conducted—not via the dispassion and detachment of bystander status, but rather via active engagement with democratic politics, especially the politics of difference. (Greene, 2002, p. 1)

Greene's evaluative work aims to democratize evaluative inquiry by privileging values and actions toward inclusion, equity, and quality in program access, experience, and outcome for all participants, especially those underserved and marginalized.

Sometimes individuals in the field would view democratic evaluation and participatory evaluation as essentially the same. While both are types of collaborative approaches to evaluation, Greene makes a notable distinction between the two approaches. She argues that in participatory approaches, there is an effort to involve the people being studied as co-evaluators, as a way of enabling them to assert their voice and views and as a capacity-building effort. This, of course, is certainly not a bad thing to do. However, Greene stresses that democratic evaluation is not a participatory process in the sense that members of the setting are primarily involved as co-evaluators. An important distinction that Greene makes is that democratic evaluation is as pluralistic and as inclusive as possible. More specifically, democratic evaluation, asserts Greene, is inclusive of stakeholder involvement in three important points of leverage in the evaluation process: setting the agenda (e.g., determining the questions and concerns to be addressed in the evaluation); ascertaining quality (i.e., including more people in the conversations about what constitutes a good program); and interpreting results and determining whatever action implications might follow them. A major emphasis of democratic evaluation that appeals to Greene is its emphasis on broad inclusion of stakeholder perspectives and the importance of power-sharing (or power redistribution) whereby local program knowledge is valued, or sometimes valued greater, than other kinds (e.g., social science-based) of knowledge.

Interfacing With Culturally Responsive Evaluation

Evaluations that explicitly aim to be more responsive to culture and cultural context represent a relatively recent phenomenon in the field, gaining prominence mostly over the past 25 years. Scholars and practitioners who espouse CRE approaches draw heavily from responsive evaluation. They see the critical need for evaluators and evaluations to be keenly attuned to and responsive toward the program itself, its larger cultural context, and the lives and experiences of the program staff, participants, and other stakeholders. Greene's work, deeply embedded in responsive and democratic evaluation approaches, seamlessly aligns with the tenets of CRE.

In 2003, I was the principal investigator for the Howard University Training Institute (HUETI), funded by the National Science Foundation (NSF). Greene graciously accepted our invitation to provide a keynote address during the inaugural summer training for mid-level evaluators interested in gaining more knowledge and skills in conducting cultural, contextually responsive, and technically sound evaluations. This was a time when CRE as an approach was being elevated within the field, and the NSF was supporting several initiatives and scholars to advance this approach, one being the HUETI.

In my opinion, it was important to have Greene deliver a keynote address to the HUETI participants on the vital topic of culturally and contextually responsive evaluation and that she aligned this approach with her own vision for evaluation. And, of course, she did not disappoint. During her keynote, Greene (2003) set the stage for this inaugural training by proclaiming:

> I believe that engagement with diversity and difference in evaluation is both substantive and a moral commitment. It is enacted in what issues we address, what methods we use, what kinds of reports we craft, and who we are as evaluators, where we position ourselves in our work, what kinds of relationships we forge with others, what we attend to and what matters in those relationships. (p. 4)

She elaborated her connections between her views and the HUETI's anchoring of its evaluation approach in the important ideas of contextual sensitivity, responsiveness, and relevance. She noted how decontextualized questions such as "How good is this program?" make no sense, but instead urged evaluators to ask questions such as "How good is this program for these children in this school and this community at this time?" And definitions of "goodness," Greene argued, themselves must become contextualized.

Today, CRE as an approach is expanding and is increasingly embraced and practiced by many in the field, including authors in this volume. It centers culture and examines processes, outcomes, and impacts through lenses in which the culture of the participants and community is considered an important factor (Frierson et al., 2010; Hood et al., 2015). Cultural dimensions are undoubtedly "front and center" in Greene's work. Consistent with CRE perspectives, Greene privileges the needs and interests of culturally marginalized groups and others who have yet to attain fair and equitable access to the resources and opportunities needed to develop their full human potential. Greene gathers evaluation information that carefully attends to the dimensions of culture and how these dimensions might show up in relationships established during the evaluation process and the data obtained.

Culture is complex; it is not a simple concept to measure or disaggregate. And Greene acknowledges the complexity of culture for the people in the programs that she evaluates. With this recognition, Greene does not try to disentangle the various entanglements of culture (e.g., race, ethnicity, and class). Instead, she respects the entanglement and represents that entanglement within her work. Greene's reflections about the various dimensions of culture in evaluation and engaging diverse stakeholders in meaningful and respectful ways are perspectives that align with CRE approaches. Additionally, as a proponent of mixed methods, Greene argues that this approach, widely advocated in CRE, advances multiple ways of making sense of evaluation data and has the potential to illuminate cultural nuances in ways not enabled with quantitative methods that either aggregate (i.e., mask) cultural differences or disaggregate them using very simplistic markers.

JUST DO IT! ENGAGING STAKEHOLDERS FROM A PERSPECTIVE OF RESPECT, HONESTY, AND TRUSTWORTHINESS

Greene argues that how evaluators conduct interactions between themselves and diverse stakeholders does matter greatly to the character, quality, and ultimate effectiveness of the evaluation. As stated earlier in this chapter, when Greene speaks of the evaluator's responsibility to enact concrete pathways for meaningful and genuine stakeholder voice and action where democratic ideas and processes are privileged, she asserts that this responsibility "requires action, and not just talk" ("Voices From the Field Interview" published in Thomas & Campbell, 2021, p. 254). This position certainly influenced Thomas and Campbell's (2021) notion of responsive stakeholder engagement, that is, engaging in an intentional and ongoing process of relationship building and connections with multiple and diverse stakeholders, paying close attention to what stakeholders are signaling as their needs (and wants) and, to the extent feasible and ethical, conducting the evaluation in a manner that is directly responsive to these needs and wants. Engaging stakeholders requires the evaluator to take tangible and explicit actions.

Greene stresses that such actions must be taken throughout the entire evaluation process. She also urges that these actions should create concrete avenues for meaningful participation by all stakeholders or their representatives. These avenues, in part, operationalize how evaluators can navigate well in others' spaces. Greene points out that avenues of demonstrating respect and gaining stakeholder trust might invoke what might be considered as non-evaluator responsibilities, such as arranging transportation or childcare for stakeholders, providing meals for participants and their families, or offering the services of an interpreter ("Voices From the Field Interview" published in Thomas & Campbell, 2021, p. 254). Democratizing inquiry for Greene entails being inclusive and providing a space for diverse stakeholders to voice their understandings, experiences, and perspectives on the program being evaluated and the social conditions that gave rise to the program. In her work, she strives toward fairness and ensuring that the experiences and stances of the least advantaged are equitably represented and respectfully heard. Further, Greene espouses the position that demonstrating respect for participants should involve, among other things, explicitly obtaining permission to share their ideas while keeping their personal stories private and honoring the commitments made to them.

VISITING AND LISTENING WELL

Greene envisions an evaluative inquiry where evaluators go out into the public world, visit with diverse others, listen well to each of them, and thereby make sound judgments in the service of doing wise actions. She views evaluation as an intrinsic part of the institutional fabric of public discourse and decision-making about public issues. Greene contends that the practice of evaluation either sanctions and reinforces, or alternatively challenges and disrupts, key dimensions of these contexts. Given this, Greene (2002) refers to evaluators as

among the weavers of this fabric [of public discourse and decision-making about public issues], not just observers or admirers or critics of the weaving. And so . . . [we should] help ensure that all the many different kinds of weavers are present and join with them in weaving a cloth resplendent with diverse yarns, multi-hued colors, and varied patterns of plaiting. (p. 5)

Greene (2002) speaks of "visiting and listening well" in our evaluation practice. Her connection between visiting and listening well in evaluation work is influenced by the perspective of Hannah Arendt (1906–1975), a 20th-century Jewish philosopher who fled Germany in 1933 and went first to France and then to the United States. In particular, Greene (2002) took note of Arendt's portrait of "good judging actors." Arendt studied the links between thinking, judging, and acting while trying to understand people, particularly trying to understand how really smart people, really good thinkers could be such bad judges and do such awful things.

Arendt focused on the interconnections between good judgment and wise action and toward developing a portrait of the "good judging actor." To be a good judging actor, according to Arendt, was not a matter of objective knowledge or subjective opinion but because of intersubjectivity. This intersubjectivity is an individual's capacity to consider other viewpoints of the same experience, that is, "to look upon the same world from another's standpoint, to see the same in very different and frequently opposing aspects" (Arendt, 1961, p. 51). Arendt (1968) advocated "visiting" or carefully listening to the perspectives of others because, in her own words, "The more people's standpoints I have present in my mind while I am pondering a given issue, and the better I can imagine how I would feel and think if I were in their place, the stronger will be the capacity for representative thinking and the more valid my final conclusions, my opinion." (p. 241). Greene argued that evaluators could and should aspire to be "good judging actors," as characterized by Hannah Arendt.

Greene stresses that evaluators must be mindful that, in their work, they are essentially a visitor in others' spaces. "Visiting" is not simply seeing through the eyes of some else, but seeing with your own eyes from a position that is not your own in a story very different from your own (Biesta, 2001). As evaluators in diverse spaces, Greene contends we must listen well to the perspectives of the different communities being represented. Listening well, as Greene (2003) describes, "means not so much to walk in another's shoes but to walk in your own shoes in someone else's story and strive to understand it (p. 18). By doing such, she asserts that we are thus enabled to make good judgments in the service of doing wise actions. And, wise actions contribute to wise practice of social programs that are strong and successful. Visiting and listening well undoubtedly call for the enactment of respect and meaningful, diverse stakeholder participation and engagement. Throughout her research and practice spanning several decades, Greene argued that stakeholder respect engenders trust, commitment, and caring, which, in turn, facilitates the possibility of engaging in more authentic conversations, collecting more meaningful evaluation data, and ultimately articulating more truthful consequences.

CONCLUSIONS

Greene has been (and continues to be) one of the most influential figures in the late 20th-century and current 21st-century in evaluation teaching, theory, and practice. She expanded and, in many respects, re-envisioned the role of evaluation and the evaluator. Over the years, Greene constantly reminds me, and many other evaluators, that good evaluation is just as much about building relationships, listening and visiting well, and creating rapport across varied stakeholder groups as it is about the technical side of crafting questions, developing evaluation designs and methods, hypothesis testing, collecting and analyzing data, and reporting results. Her beliefs, evaluative thinking, and evaluation practice were heavily influenced by her parents, upbringing, progressive sensibilities, and "coming of age" during one of the most pivotal times in this country's focus on equity and civil rights.

At present, when our society is once again revisiting issues of systemic racism, justice, and diversity, Greene continues to serve as a beacon for promoting equity and engagement in evaluation and the broader society. And, rightfully so, she has been honored by various universities (both nationally and internationally), professional organizations (e.g., American Evaluation Association, American Educational Research Association), and numerous colleagues and students whom she mentored and "uplifted."

Greene's work and significant presence have steadfastly moved the field of evaluation toward positioning its scholarship and practice more deliberately in service of the public good. Her gentle spirit, willingness to share her expertise, and consequential collaborations with diverse scholars and practitioners, including me, are "living" testaments to her commitment to and passion for advancing democratic and inclusive approaches in evaluation and toward expanding diversity (e.g., methodologically, racially, culturally) within the field. Greene's democratic and inclusive approach to evaluation, in general, and her ability to visit and listen well, more specifically, are important parts of her evaluation practice and are probably her greatest legacy. And, as for me, I am forever indebted to her for the inspiration and support over two decades.

REFERENCES

Arendt, H. (1968). *Between past and future*. Penguin.

AZ Quotes. (n.d.). Isaac Watts quote. Retrieved from https://www.azquotes.com/quote/1157023

Frierson, H. T., Hood, S., Hughes, G., & Thomas, V. G. (2010). *A guide to conducting culturally responsive evaluations. In the 2010 user-friendly handbook for project evaluation* (pp. 75–96). NSF.

Greene, J. C. (2002, October/November). *Toward evaluation as a "public craft" and evaluators as stewards of the public good or listening well* [Keynote address]. 2002 Australasian Evaluation Society International Conference, Wollongong, Australia. Retrieved from http://citeseerx.ist.psu.edu/viewdoc/download?doi=1 0.1.1.517.770&rep=rep1&type=pdf

Greene, J. C. (2003, July 16). *Evaluation as engagement with difference: In service of the public good.* Keynote address delivered to the Howard University Evaluation Training Institute. Washington, DC.

Greene, J. C. (2005a). Evaluators as stewards of the public good. In S. Hood, R. Hopson, & H. Frierson (Eds.), *The role of culture and cultural context: A mandate for inclusion, the discovery of truth and understanding in evaluative theory and practice* (pp. 7–20). Information Age Publishing.

Greene, J. C. (2005b). A value-engaged approach for evaluating the Bunche-Da Vinci Learning Academy. *New Directions for Evaluation, 2005*(106), 27–45. https://doi.org/10.1002/ev.150

Greene, J. C. (2015). Evaluation as a socio-political intervention. *Spazio Filosofico,* 13, 87–95. Retrieved from https://www.spaziofilosofico.it/wp-content/uploads/2015/02/Greene.pdf

Greene, J. C. (2016, October 12). Valuing values in evaluation. *CREA in the 21st century: The New Frontier.* Retrieved from http://creablog.weebly.com/blog/valuing-values-in-evaluation#:~:text=Evaluation%20is%20the%20social%20practice,intervention%20somewhere%20in%20the%20world

Greene, J. C. (2018a). In gratitude to Jim and Whit. *New Directions for Evaluation, 2018*(157), 74–77. https://doi.org/10.1002/ev.20291

Greene, J. C. (2018b). Jennifer Greene's one top YEE tip. *Evaluation for Development.* Retrieved from https://zendaofir.com/top-tip-yees-jennifer-greene/

Greene, J. C., Boyce, A. S., & Ahn, J. (2011). *A values-engaged, educative approach for evaluating education programs: A guidebook for practice.* University of Illinois.

Greene, J. C., DeStefano, L., Burgon, H., & Hall, J. (2006). An educative, values-engaged approach to evaluating STEM educational programs. *New Directions for Evaluation, 2006*(109), 53–57. https://doi.org/10.1002/ev.178

Hall, J. N., Ahn, J., & Greene, J. C. (2012). Values engagement in evaluation: Ideas, illustrations, and implications. *American Journal of Evaluation, 33*(2), 195–207. https://doi.org/10.1177/1098214011422592

Hood, S., Hopson, R., & Frierson, H. (Eds.). (2015). *Continuing the journey to reposition culture and cultural context in evaluation theory and practice.* Information Age Publishing.

House, E. R. (2014). *Evaluating: Values, biases, and practical wisdom.* Information Age Publishing.

House, E. R., & Howe, K. (1999). *Values in evaluation and social research.* SAGE Publications.

Tarsilla, M. (2010). Theorists' theories of evaluation: A conversation with Jennifer Greene. *Journal of MultiDisciplinary Evaluation, 6*(13), 209–219.

Thomas, V. G., & Campbell, P. B. (2021). *Evaluation in today's world: Respecting diversity, improving quality and promoting usability.* SAGE.

Thomas, V. G., & Stevens, F. I. (Eds.). (2004). *Co-constructing a contextually responsive evaluation framework: The Talent Development Model of School Reform.* Jossey-Bass.

University of Illinois Urbana-Champaign. (2017, May). Center for Innovation in Teaching & Learning: I-STEM Education Initiative. https://www.istem.illinois.edu/news/diversity.panel.html

Worthen, B. R. (2004). Whither evaluation? That all depends. *American Journal of Evaluation, 22*(3), 409–418. https://doi.org/10.1177/109821400102200319

CHAPTER 7

GOOD START

Creating Change From the Ground Up

Sharon Brisolara
Inquiry That Matters

Tristi Nichols
Manitou.Inc

Kathryn A. Sielbeck-Mathes
Measurement Matters-Personalized Evaluation
& Consulting Services, LLC

Denise Seigart
East Stroudsburg University

ABSTRACT

What are our reflections, four senior evaluation professionals of diverse backgrounds, regarding Jennifer Greene's mentorship, scholarship, wisdom, and evaluation practices? We completed our doctoral work at Cornell University under Greene's guidance and commonly heard, "good start" as a way to com-

Disrupting Program Evaluation and Mixed Methods Research for a More Just Society, pages 95–106
Copyright © 2023 by Information Age Publishing
www.infoagepub.com
95

municate that we were on the right path. In this chapter, we describe how Greene's framing of evaluative thinking was foundational and critical to our emerging conceptualization of equity, social justice, and democratic values. Our chapter adds to the limited literature on values underpinning evaluation approaches with narratives about the power of enacting those values and the ways in which enacting values requires commitment, must be nurtured, and demands ongoing reflection. In addition to attending to transparency and truth telling, we draw attention to the need to always have respect for the help or harm that evaluation can do, as there are real consequences for the people most impacted and at the center of our enterprise.

Good Start. One of us remembers these words as an embodiment of Jennifer's response to assignments and dissertation chapters. As doctoral candidates at Cornell University under Jennifer's tutelage during the mid-1990s, a good start is also what the authors received in conversation with Dr. Greene. In our chapter, we examine Jennifer Greene's lifelong work from the vantage point of four evaluation professionals, working in various fields, who have integrated her teaching and scholarship on values-grounded inquiry into our work in somewhat different ways. As educators, evaluators, and professionals who have brought evaluative inquiry into our work, we have been significantly shaped by our initial orientation toward the core commitment to democratizing inquiry evident in Jennifer's writing and practice. Reflecting on our experience of Jennifer as evaluation scholar, educator, mentor, and friend, we focus on what we learned about genuine fostering stakeholder engagement, committing to individual and collective learning in service of equitable outcomes, and cultivating deep attention to the creation of inclusive and culturally responsive practices. In so doing, we reflect on her consistent commitment to values-based, evaluative thinking; insistence on engaged participation; her modeling of authentic relationships; and the centrality of personal and collective learning in the project of disciplinary transformation.

We begin with our grounding in evaluation and evaluative theory in the early 1990s, a reminder of Jennifer's critical role in shaping the transformations to come and the disruption required at the time. With this background in mind, we revisit how Jennifer's framing of evaluative thinking was foundational to the emerging conceptualization of equity as central to culturally responsive evaluation. Our experience with Jennifer cemented for us the notion that beyond what typically passes for inclusion and collaboration lies authentic relationship and care, a way of being together that is currently reflected in the centrality of self-care and community care in supporting and sustaining social justice efforts. A focus on equity rooted in authenticity and care promotes a clarity of vision, both inward and outward, witnessing and offering what is seen with such a lens to an interpretation process that benefits from a range of perspectives. We end with another good start, acknowledging that democratizing inquiry is an ongoing project, one that requires true inclusion, humility, authentic care, curiosity with an intent to learn, community, and commitment.

OUR GROUNDING: A GOOD START

One day you finally knew what you had to do and began. Though the voices around you kept shouting their bad advice—though the whole house began to tremble, and you felt the old tug on at your ankles.
—Mary Oliver (Oliver, 2023), "The Journey"

Every journey begins somewhere and has left somewhere else. We each came to Cornell and Jennifer with our strengths and our doubts, our challenges and our gifts. We encountered an evaluation field that has since been described as being primarily guided by research methodology (Shadish et al., 1991). There were essentially two paths of study from which to choose: post-positivism and constructivism. Each paradigmatic choice was thought to have a direct effect and ostensibly long-term impact on one's evaluation practice, including where one could work, with whom, and for what purpose(s).

The positivist paradigm, the bedrock of experimental and quasi-experimental designs, relied on treatment and control groups in search of causality or its approximation. Students drawn to this paradigm learned related research design options and immersed themselves in learning "threats to validity" all of which has been well documented in the evaluation literature (Rossi & Freeman, 2004, p. 515). The constructivist perspective was garnering attention when we began our studies and emphasized a revisioning of core aspects of knowledge construction: ontology, the study of knowledge; epistemology, the relationship between the knower and known, the researcher and researched, the evaluator and stakeholders; and methodology or the conduct of inquiry (Guba & Lincoln, 1989). We were introduced to interpretivist paradigms that grew out of phenomenology and hermeneutics. Discussions and debates were ubiquitous as "increasing questions emerged about the focus of inquiry, as well as exploration of methodologies that emphasized discovery, description and meaning rather than prediction, control and measurement" (Laverty, 2003, p. 21). Through it all, we studied the various pillars of the evaluation field, some of whom appear in this volume.

Jennifer has reflected beautifully on this period:

Those were heady times—exceptionally intellectually demanding, as many of us were challenged to learn a whole host of new concepts and ideas, many of them highly abstract and deep. We read widely, eager for any books or articles that could help us learn as we strove to be among the enlightened. We jammed into conference sessions that featured people in the know. We met in revolt and in protest against revolt, established new interest groups and convening small conferences to advance our disparate interests. Heroes and heroines emerged, as did anti-heros and anti-heroines. Which was which obviously depended on your point of view. (Greene, 2007, p. 41)

It was, indeed, an exciting time and somewhat daunting to those of us beginning our foray into evaluation amidst vigorous debates for what felt like the soul of the profession. Was truth a singular entity on which one could expect to achieve consensus, multiplistic and reflective of different voices and agendas, co-constructed/created or unearthed from a joint reality (Glaser & Strauss, 1967; Strauss & Corbin, 1990)? Must one attempt to control confounding variables or explore a constructed reality where dynamic truths emerge (Guba & Lincoln, 1989)? What constituted rigor? Who can be legitimately engaged in knowledge production? The approaches above offered very different roles for the evaluator, different ways of knowing (e.g., "objective" detached or embedded "subjective"), and correspondent methodological commitments given the understanding that a choice (quantitative or qualitative) was required being so aligned with different paradigms that were thought by some to be incompatible with each other.

As Jennifer's students, we were drawn into these debates and the then groundbreaking philosophical discussions around mixed-method and participatory approaches, democratic evaluation, and emerging efforts to establish feminist evaluation. We engaged and eventually came together to collaborate on a *New Directions for Evaluation* volume titled "Feminist Evaluation, Explorations and Experiences" as a way to be in service of the core commitments we had made and honor the work of those who had spent so many years laying the groundwork for this model. Through it all, we grappled with understanding and making, as transparent as possible, our own subjectivity, intersubjectivity, and meaning making through reflexive processes. We struggled with what it meant to be an evaluator, to understand ourselves and our diverse backgrounds within these discussions, and to be ethical and honest in the different cultures and contexts we intended to work. It was not until a few years later that one could access practical discussions and literature describing how evaluation practitioners responded to cultural context in their practice and how culture was conceptualized and defined while making evaluative judgments (Chouinard & Hopson, 2015). And yet, the tenor of the times contributed to our cultivating awareness of the necessity to recognize and acknowledge the cultural self that accompanies us into the field (Scheper-Hughes, 1992).

EQUITY AND EVALUATIVE THINKING: FIRST STEPS

Research is formalized curiosity. It is poking and prying with a purpose.
—Zora Neale Hurston (Hurston, 1996, p. 143)

As evaluation practitioners and researchers today, we continue to draw on Greene's scholarship and mentorship. Both were invaluable in forming our capacity to use evaluative thinking in ways and through frameworks that honor and seek to advance the democratic ideals and equity. There are many definitions of equity. Common to most is a distinction from equality, attention to access, opportunities for meaningful participation, and equitable positive outcomes through particular attention to those marginalized and least well-served by existing structures and dynamics. One way of thinking about equity is just and fair inclusion into a society where all can participate, prosper, and reach their full potential.

While the word "equity" was not commonly used at the time to describe the intent of democratizing inquiry, it was central to our discussions of how to enact social justice, meaningful inclusion, and democratic participation in our evaluative work. Jennifer's insistence on evaluative thinking and ethical reflection contributed to our developing clear positions on the approaches we use. Those positions have been informed by the values we hold and the belief that evaluation, its process, and outcomes can be a powerful resource for making a positive difference in people's lives. This cannot happen without attention and intention. To attend and tend carefully to equity reflects that commitment in the types of evaluation questions we ask, the methods we use, the data we collect, the analysis and interpretation strategies we select, and the ways we share results. It means ongoing engagement in the question of the extent to which the evaluation is truly serving those most significantly affected by the program and whether the results and process meaningfully contribute to ameliorating the social problem being addressed. It requires deep commitments to ensuring that those approaches are assessed through equity and social justice frameworks.

One of the ways that equity concerns manifested for us was our work on the aforementioned volume on feminist evaluation. Our research, experiences, and identities had drawn us to understand the need for a lens that began with gender and acknowledged the diverse ways in which people, situations, and programs were situated in dynamics of power, at the intersection of gender, class, race, and other identities, and engaged in diverse forms of knowledge construction and consumption. Jennifer was instrumental in our connecting with a group of established evaluators who had spent years advocating for a feminist evaluation approach. They had encountered ongoing resistance to the publication of writing on this model and decided that it might be time to release the project into other hands.

They shared their previous efforts and challenges, their sense of the need for the work, and their encouragement to make the work our own.

For us, the work was driven by a desire to articulate what was not being seen, including attention to the social structures and systems that give rise to discrimination and oppression. Values of justice, inclusion, and equity (e.g., gender, racial, and more) pushed the work forward. The time was right, and the groundwork had been laid; greater openness to the approach led to publication and further dissemination of such discussions. The experience taught us valuable lessons about the power of enacting values, of doing so in community, and of the reality that a commitment to equity is a lifelong journey that will be tested, must be nurtured, and requires responsiveness and ongoing reflection.

BUILDING CHANGE CAPACITY: THE POWER
OF AUTHENTIC CARE

My work is about the establishment of trust. For someone to share their authenticity with me is a soul-to-soul thing. It's not a lens-to-soul thing.

—Photographer Lisa Kristine (Chakraborty, n.d.)

As is likely evident in this chapter, our engagement with Jennifer's scholarship was strongly linked to our engagement with Jennifer, the educator, evaluator, and deeply human being. Indeed, the power of her work, for those of us in academia, lies in how fully she embodied the values she claimed and shared. Her conceptualization of evaluative thinking embraced developing a mindset grounded in critical thought, attentive to systems and values clarification, and based on authentic interaction. She has encouraged evaluation practitioners, regardless of one's practice context, "to name and claim the values that are advanced in the evaluation work that we do" and to nurture an evaluation habit where "there is plenty of space and potential support for a central emphasis on equity, for the argument that at least one of the values that should be advanced in our evaluation practice is that of equity" (Greene, 2016, p. 55). This is a call for transparency and accountability for the purpose of deeper dialogue about how decisions should be made to advance equity and the social good. Through our work on feminist evaluation and our efforts in different fields, we have come to espouse the view that transparency of value commitments and intents and authentic representation continue to demand our focused attention and examined clarity. We have also come to understand that transparency, accountability, and clarity require a level of deep engagement if we are to learn about hidden dynamics and our responsibilities in reinforcing or deconstructing them. We benefit from entering such spaces mindful, open, and responsive, rooted in an ethic of authentic care.

Equity and justice, democratic values, can only flourish where relationships are authentic and grounded in trust, safety, and inclusion. Transparency and truth-telling are important, but it is the way in which we treat each other, the extent to which we care, that matters. Jennifer's "instructional practice" was personal, moving beyond andragogy and curricula to a way of interacting with learners, a reflection of an ethic of authentic care or Davis and Kaminsky's notion earlier in this volume of "holding space." Such authentic, caring relationships are essential for evaluators, particularly those who are not longstanding members of the communities they are evaluating, dedicated to co-creating the conditions in which underlying issues can rise to the surface. Authentic caring relationships provide the foundation from which people can collectively speak to the role of systems and structures, power dynamics, ethical breaches, barriers to participation, unintended benefits, and the nontangible rewards of participation. Respect and care, or the absence of either, will shape how and to what extent all affected by current programs, policies, and practices engage in the co-construction and use of knowledge.

Within the field of education, scholars such as Strayhorn, Wood, and Harris have made strong cases for the centrality of safety and an ethic of care as a precursor to educational attainment. Their work highlights the particular importance of care for those from marginalized populations with years of experience interacting with societal institutions permeated with White supremacist ideologies (Sadowski, 2016; Strayhorn, 2012; Verschedlen, 2017; Wood & Harris, 2016). Engagement with others, often conceptualized as a form of relationship, turns out to be how we learn (National Academies of Sciences, Engineering, and Medicine 2018). Even in individualistic cultures, engagement, interaction, and relationships are essential to taking in information for the broader goal of learning. Learning can be thought of as a community practice (Wagner, 2011) with engagement existing on a continuum. This understanding was reflected in democratic evaluation's dedication to ensuring that people had access to information about an evaluation to effectively engage in debate about the issues at stake (Hanberger, 2006). Similarly, responsive evaluation drew on hermeneutic traditions that advanced the importance of dialogue for individual and collective understanding. "It is," as Abma et al. (2020) have stated, "concerned with moral learning and relational responsibilities" (p. 131).

As dissertation chair, Jennifer often encouraged her students to connect with other students to deepen their learning, a nudge that led to one author's involvement in a dissertation support group that resulted in much laughter, a few tears, and the courage to reflect more deeply on what truly mattered about one's research and work. Jennifer encouraged us as young evaluators to seek community and dialogue through small conferences and at the American Evaluation Association (AEA) annual meeting for the purpose of engaging in intellectual debate within a supportive but contentious community of truth seekers. Doing so requires courage, vulnerability, and trust, all of which are scaffolded by community. We have experienced community and care over many years in witnessing her investment in her doctoral students (past, current, and future), in the midst of all of the demands on her time, and the generosity embedded in her dedication to connecting people. Her actions made clear what engagement entailed. It was a lesson in the value and lodestar of community we come back to again and again: to never lose sight of the purpose of evaluation, to have deep respect for the good or harm it can do, and to ground ourselves in the very real consequences for the people at the center of our enterprise.

Feminist and womanist theories reintroduced to the academy reflections on how emotions, love, and empathy operate as legitimate sources of knowledge and ways of knowing (see, e.g., Jagger, 1989). In recent years, the Black Lives Matter movement and others seeking transformative change have focused attention on new constructs of self-care and its relationship with community care in sustaining social justice work. In these contexts,

> *Care* refers to a relational set of discourses and practices between people, environments, and objects...Theorized as an affective connective tissue between an inner self and an outer world, care constitutes a feeling with, rather than a feeling for, others. When mobilized, it offers visceral, material, and emotional heft to acts of preservation that span a breadth of localities: selves, communities, and social worlds. (Hi'ilei & Kneese, 2020, p. 2)

The emphasis of care here is on the needs of those most harmed by the systems currently in place.

An ethic of authentic care makes engaged participation possible. Jennifer modeled engaged, authentic care in her support of a (new to the discipline) participatory evaluation dissertation proposal to be conducted in Central America. In relationship with one of the authors, she consistently brought deep questions that explored the meaning and possibilities of participatory evaluation, the continuum of participation represented by diverse collaborative models, and reflections on standards for engaging in cross-cultural evaluation. It was authentic care that allowed us all to grapple more deeply with criteria for determining a program's readiness for feminist evaluation and clarity about ways of presenting, representing, and talking about the approach. Looking around at the challenges in discourse and understanding, even of key terms, that seem to surround many of our communities in the United States today, we see a lack of experience of authentic care that so often confounds our ability to build capacity for change.

How prescient Jennifer's call to firmly ground our work in such values and relationships seems today. It is too easy to despair about the fragile and contingent nature of our attempts to speak to each other about values; it is painfully clear how vulnerable and contested evidence is absent of engagement, dialogue, and relationship. Jennifer's great gift, in the way her work and her life align, is in bringing home the essential role, and often the life-and-death nature of how we do what we do, how we nurture and compose the evaluative habit, and the rigor and vulnerability with which we examine ourselves as an evaluative instrument.

ADJUSTING OUR GAZE: WITNESSING AND PRESENCE

Bearing witness, like solidarity and compassion is a term worth rehabilitating. It captures ways of knowing, both forms of silence. Bearing witness is done on behalf of others for their sake (even if those others are dead or forgotten). It needs to be done, but there is no point in exaggerating the importance of the deed. (Farmer, 2005, p. 28)

Being with Jennifer was transformative in its insistence to draw deeply not only from a range of evaluation scholars, but also from other fields and literature resources, particularly given our diverse backgrounds and existing experience in other fields. This stance was consistent with what we viewed as her attention to a paradigmatic agnosticism (our term) and rich appreciation of diverse perspectives, disciplinary tools, and sources of knowledge. And this stance was, in many ways, deeply pragmatic. She said, "I believe that conscious attention to how the various strands of mental models influence inquiry decisions renders such decisions more thoughtful, reflective, intentional, and thereby more generative and defensible" (Greene, 2007, p. 59). By the end of our doctoral studies, we saw "no meaningful epistemological difference between qualitative and quantitative methods. Instead, we [saw] both as assisted sensemaking techniques that have specific benefits and limitations" (Mark et al., 2000, p. 16). It was in Jennifer's pioneering article "Crafting Mixed-Method Evaluation Designs" that this was labeled the pragmatic position (Greene & Caracelli, 1997). Later, many theorists made peace with mixing "opposing" paradigms, and we embraced strategies for describing our hard-won positions.

We adjusted our gaze again and again as our understanding grew. The more deeply we researched the possibilities of feminist evaluation in the context of the contributions of participatory, democratic, responsive, empowerment, and other challenger/transformative approaches, the more we understood the critical role of self-awareness in our work. Current social justice activists might use different words to explain this imperative—that knowing ourselves and our biases, learning edges, weaknesses, and strengths is essential to further learning, the right relationship, and the ability to see. We came to evaluation seeking particular tools to do meaningful work from different backgrounds and levels of confidence. One author remembers herself as a young nurse and academic who knew she needed a PhD to advance in her career and stumbled into the program at Cornell because it seemed like a better fit than any other doctoral program she had examined. Like many of her peers, she often thought, "How the hell did I get into this program?" She and others wondered what Jennifer was thinking when she accepted the role of dissertation committees chair. As the years went by, the author fell in love with the field of evaluation, disappointed that she had never been challenged before the way she was with Jennifer at Cornell.

We have come by way of gratitude to the knowledge that we all need a Jennifer: educator, facilitator, coach, agitator, social justice advocate, compassionate judge, and friend. We all need a partner and guide who can help us question everything, walk with a *beginner's mind*, see through internal and external jumbles to better understand the whole, and work for the greater good that Symonette describes in an earlier chapter. With the gifts we have been given, we can also be partners to others seeking to understand the meaning from particulars. We can be present and engaged, like the poet Mary Oliver describes about her experience taking in the shell-strewn edge of the sea in her poem "Breakage." "It's like a schoolhouse/of little words,/ thousands of words./First you figure out what each one means by itself,/ the jingle, the periwinkle, the scallop/full of moonlight./Then you begin, slowly, to read the whole story" (Oliver, 2003, n.p.).

CONTINUING TO RISE: ONE GOOD START AFTER ANOTHER

If we did not know from which paradigm we were functioning or what commitments we most cherished when we entered study with Jennifer, we certainly did by the end of our time at Cornell. Our experience with Jennifer and her work has been meaningful and transformative for us as individuals interacting in a field that, Bledsoe reminds us, has been deeply shaped by her presence. Jennifer invited us to commit to democratizing inquiry, not as an approach or model, but as a framework and mindset. This perspective would guide the ways we understood and created the conditions for inclusion, the ways we acknowledged the partiality of our perspectives and engaged in collaborative efforts to learn, and the ways we contributed in our own ways, regardless of what we did, to justice, equity, and good. Wherever our paths have taken us and whatever our foci, we journey clear about our commitments to the people our work will affect and serve, the effects of our practice and ways of relating, and what is required of us in caring about the particular public good, from start to finish and beyond.

REFERENCES

Abma, T. A., Visse, M., Simons, H., & Greene, J. C. (2020). Enriching evaluation practice through care ethics. *Evaluation, 26*(2), 131–146. https://doi .org/10.1177/1356389019893402

Caracelli, V. J., & Greene, J. C. (1997). Crafting mixed-method evaluation designs. *New Directions for Evaluation, 1997*(74), 19–32. https://doi.org/10.1002/ev.1069

Chakraborty, D. (2019, May). Humanitarian photographer Lisa Kristine documents the world. *Professional Photographer.* https://www.ppa.com/ppmag/articles/ humanitarian-photographer-lisa-kristine-documents-the-world

Chouinard, J. A., & Hopson, R. (2015). A critical exploration of culture in international development evaluation. *Canadian Journal of Program Evaluation, 30*(3), 248–276. https://doi.org/10.3138/cjpe.30.3.02

Farmer, P. (2005). *Pathologies of power: Health, human rights, and the new war on the poor.* University of California Press.

Glaser, B. G., & Strauss, A. L. (1967). *The discovery of grounded theory. Strategies for qualitative research.* Aldine.

Greene, J. C. (2007). *Mixed methods in social inquiry.* Jossey-Bass.

Greene, J. C. (2016). Advancing equity, cultivating an evaluation habit. In S. I Donaldson & R. Picciotto (Eds.), *Evaluation for an equitable society* (pp. 49–66). Information Age Publishing.

Greene, J. C., & Caracelli, V. J. (1997). Advances in mixed-method evaluation: The challenges and benefits of integrating diverse paradigms. Jossey-Bass.

Guba, E. G., & Lincoln, Y. S. (1989). *Fourth generation evaluation.* SAGE Publications.

Hanberger, A. (2006). Evaluation of and for democracy. *Evaluation, 12*(1), 17–37. https://doi.org/10.1177/1356389006064194

Hi'ilei, J. K., & Kneese, T. (2020). Radical care: Survival strategies for uncertain times. *Social Text, 38*(1), 1–16. https://doi.org/10.1215/01642472-7971067

Hurston, Z. N. (1942). *Dust tracks on a road.* J. B. Lippincott. [Review of Dust Tracks on a Road]. Harper Perennial.

Jaggar, A. M. (1989). Love and knowledge: Emotion in feminist epistemology. *Inquiry, 32*(2), 151–176. https://doi.org/10.1080/00201748908602185

Jensenius, A. R. (2012, March 12). *Disciplinarities: Intra, cross, multi, inter, trans.* https://www.arj.no/2012/03/12/disciplinarities-2/

Kristine, L. (n.d.). https://Www.mercurynews.com/2013/02/07/Moraga-Lisa -Kristine-Shares-Insights-About-Where-Photography-Humanitarian-Causes -Meet

Laverty, S. M. (2003). Hermeneutic phenomenology and phenomenology: A comparison of historical and methodological considerations. *International Journal of Qualitative Methods,* September, *2*(3), 21–35. https://doi.org/ 10.1177/160940690300200303

Mark, M. M., Henry, G. T., & Julnes, G. (2000). *Evaluation: An integrated framework for understanding, guiding, and improving policies and programs.* Jossey-Bass.

National Academies of Sciences, Engineering, and Medicine. (2018). *How people learn II*. National Academies Press. https://doi.org/10.17226/24783

Oliver, M. (2003). *Breakage*. The Poetry Foundation. Retrieved from https://www.poetryfoundation.org/poetrymagazine/poems/41917/breakage

Rossi, P. H., & Freeman, H. E. (1993). *Evaluation: A systematic approach* (7th ed.). SAGE Publications.

Sadowski, M. (2016). *Safe is not enough: Better schools for LGBTQ students*. Harvard Education Press.

Scheper-Hughes, N. (1992). *Death without weeping: The violence of everyday life in Brazil*. University of California Press.

Shadish, W. R., Cook, T. D., & Leviton, L. L. (1991). *Foundations of program evaluation: Theories of practice*. SAGE Publications.

Strauss, A. L., & Corbin, J. M. (1990). *Basics of qualitative research: Grounded theory procedures and techniques*. SAGE Publications.

Strayhorn, T. (2012). *College students' sense of belonging: A key to educational success for all students* (1st ed.). Routledge.

Verschedlen, C. (2017). *Bandwidth recovery: Helping students reclaim cognitive resources lost to poverty, racism and social marginalization*. Stylus Publishing.

Wagner, E., Traynet, B., & de Laat, M. (2011). *Promoting and accessing value creation in communities and networks: A conceptual framework*. Rapport 18. Ruud de Moor Centrum, Open Universiteit.

Wood, J. L., & Harris III, F. (2016). Supporting men of color in the community college: A guidebook. Montezuma Publishing.

CHAPTER 8

EVALUATOR'S COMMITMENT TO EPISTEMIC JUSTICE AND THE NEED OF ETHICS WORK

Tineke A. Abma
Leiden University Medical Centre

Susan Woelders
Leiden University Medical Centre

ABSTRACT

In line with Jennifer Greene's ideas on the role of evaluation as advocacy, this chapter explores the evaluator's moral responsibilities and "ethics work" when embracing the values of epistemic justice, social inclusion, and genuine stakeholder participation in evaluation. Based on our experiences in the field of psychiatry, we illustrate that the democratic evaluator has to deal with power. Epistemic injustice may hinder specific stakeholders from being acknowledged in their capacity as knowers and reproduce their exclusion to the process of knowledge production. Creating space for the voices and values less heard and taking a stance when it comes to the genuine participation and inclusion of all involved may take the courage of the evaluator. Foucault's concept of parrhesia captures the virtue of bringing an issue of social or epis-

Disrupting Program Evaluation and Mixed Methods Research for a More Just Society, pages 107–120
Copyright © 2023 by Information Age Publishing
www.infoagepub.com
107

temic injustice to the table and questioning well-rehearsed knowings, taken-for-granted notions, habits, and routines. Bringing up these issues can be seen as parrhesiastic resistance against dominant relations of power. Greene's explicit value engagement has encouraged us to take an interest and stand for social justice, democratic values, and equal participation of all, and to become advocates, even if the context expects us to be neutral and value-free.

Jennifer Greene can be considered the grand dame of democratic evaluation. Her integrity, friendship, and work have been an enduring source of joy and inspiration for many scholars and practitioners, including us. This chapter is written in honor of her personality and ideas, particularly her outspoken call for advocacy in evaluation and frank openness about one's value engagement as an evaluator. In 1997, Greene showed a lot of courage to state that every evaluation is value-laden and that evaluators should therefore not hide behind a mask of value neutrality, but instead be open about the values and voices they want to promote in their practice. "The important question," according to Greene (1997), "then becomes not, should we or should we not advocate in our role as evaluators, but rather what and whom should we advocate for?" (p. 2). This question still triggers the evaluation community after so many years because it contradicts well-rehearsed notions about evaluation as an impartial, value-neutral, and objective practice. Greene, however, withstood critics and remained an advocate of democratic pluralism, equality, equity, and social justice throughout her career. This was not just a matter of scholarly and intellectual debate. Greene embodied and lived by these values and actively brought them into practice. This made her not only a respected scholar but also a great person and teacher.

Our chapter is part of the third section on democratizing inquiry. This section aims to make explicit the ways in which Greene *democratized* inquiry by emphasizing inclusive, collaborative, culturally responsive, and educative inquiry stances and practices. The editors have conceptualized a democratic orientation toward inquiry and evaluation broadly as practices that use inclusive, collaborative, and culturally responsive and educative practices to mitigate inequalities. This includes the promotion of genuine and active stakeholder participation where issues of democratic ideas and processes are privileged. This chapter specifically focuses on the role of power in organizations when one as an evaluator intends to advocate for the participation of all stakeholders, including those whose voices are less heard. The case example we rely on in this chapter is in the field of psychiatry. Although it goes a while back, the patterns encountered when democratizing a practice are still relevant today; we still encounter the same issues and challenges in our current evaluation practices. We will explore and try to capture our experiences as evaluators who try to democratize the field of psychiatry through evaluation and reflect on the political pressure and resistance as well as the moral responsibilities and ethics work that one may encounter. Core theoretical notions that guide our reflection include Fricker's epistemic injustice and Foucault's parrhesia. We start with a presentation of our vision of evaluation as a "juggling act" and continue with the argument that moral learning takes place in the "swampy lowlands" of the evaluation practice. The heart of our chapter is built around the section called "recognizing one's responsibility," wherein we will deal with four responsibilities of the democratic evaluator: (a) gaining confidence and building trustful relationships, (b) opening spaces for the voices of people who are less able to express themselves, (c) resisting external pressure, and (d) acknowledging one's power and non-neutrality. The conclusion reiterates Greene's contribution and thinking about democratic evaluation as a value-laden practice, emerging in a power-laden context.

EVALUATION AS A JUGGLING ACT

Democratic, participatory, and responsive evaluation approaches are grounded in democratic ideals and values, notably social justice and inclusion, democratic decision-making, mutual learning, and collective action (Greene 2006, 2010; Hanberger, 2006). One of the fundamental principles in Jennifer Greene's work on democratic evaluation is the commitment to epistemic justice (Simons & Greene, 2018). Epistemic justice means that *all* whose life or work is at stake should be enabled to contribute and influence the process of knowledge production, including processes of evaluation, especially those who are marginalized (Fricker, 2007). Their inclusion and participation in the evaluation are needed to give them a say in processes affecting their lives and work and a more truthful understanding of our complex world. These multiple perspectives should be part of a dialogue and deliberation to create a situation wherein the exchange of perspectives and mutual learning can occur among all involved. Dialogue is a special form of communication grounded in respect, openness, and a "trained capacity for otherness" (Gadamer, 1960, p. 10). In dialogue, people do not seek to establish and reproduce well-rehearsed notions (Cook, 2021). On the contrary, in dialogue, people search for new meanings and mutual understanding. In this process, perspectives can merge into a "fusion of horizons" and the emergence of new territories, collaborations, and practice improvement.

Dialogue and deliberation are ideal speech situations or what philosopher Jürgen Habermas (1987) would call *Herrschaftsfreie Kommunikation.* In his critical philosophy of modern societies, Habermas made a distinction between strategic communication wherein parties try to realize their own vested power and interests (system), and communicative action where people meet each other as human beings and try to understand each other (lifeworld). According to Habermas, there is a lack of communicative action and space in modern societies, and we need communicative spaces where ideal speech situations can emerge. In an ideal situation, power asymmetry and hierarchy are temporarily put aside, and people involved are willing to investigate all claims put to the fore. Everyone's arguments are considered valid and treated equally. Habermas's ideal is worth striving for and has inspired the democratic evaluation approach laid out by Ernest House, the mentor of Jennifer Greene, and has inspired her work and ours as well. It means that the evaluator is more than just a judge or expert. A democratic deliberative evaluator is responsible for engaging all voices by creating a communicative space and staying alert to the exclusion of voices. Simons and Greene (2018) characterize this work as a "juggling act" (p. 89). The juggling metaphor nicely captures the complexity of the work completed by the democratic evaluator. The ideal speech situation may be worth striving for but does not consider that in real life, the evaluator has to deal with power and interests, both in terms of the power asymmetries in the context and the power the evaluator self holds. Indeed, evaluation is a juggling act because it occurs in the "interference zone" between system and lifeworld (Abma et al., 2016; Woelders & Abma, 2017). For example, in policy making and evaluation, the standards and knowledge of experts and decision makers are often taken for granted. This may not equally be the case for people it concerns and whose interests are at stake. People may experience a more challenging time being taken seriously and bringing in their lifeworld perspective, which does not fit the standards. A situation Miranda Fricker (2007) would denounce as "epistemic injustice" (p. 1). Fricker argues:

> In all such injustices the subject is wronged in her capacity as a knower. To be wronged in one's capacity as a knower is to be wronged in a capacity essential to human value. (p. 5)

Testimonial justice occurs when a speaker is given less credibility because of the prejudice of the hearer. This implies that some people are not afforded the opportunity to be heard. Fricker (2007) also adds another form of injustice to knowledge and knowledge production. A subject also receives and builds knowledge out of social interaction by becoming part of a social experience. Excluding people from activities in which they can learn can lead to hermeneutic injustice:

> If one lives in a society or a subculture in which the mere fact of an intuitive or an emotional expressive style means that one cannot be heard as fully rational, then one is thereby unjustly afflicted by a hermeneutical gap—one is subject to a hermeneutical injustice. (p. 161)

Ethics, power, and knowledge are intertwined in Fricker's concept. If the experiences and voices of some are systematically questioned while others are accepted and taken for granted, the question raises what this means for the democratic evaluator in terms of responsibilities. In this chapter, we want to reflect on these responsibilities of the democratic evaluator. An important notion is the idea that the evaluator should perform "ethics work" to live up to the normative ideals of democratic and related evaluation approaches (Abma, 2020; Abma et al., 2020; Groot, 2021). Ethics work entails the effort one puts into seeing ethically salient aspects of situations, developing oneself as a reflexive practitioner, paying attention to emotions, collaboratively working out the right course of action, and participatory reflection in the company of critical friends. Such ethics work is part of everything we do or not do, how we interact with others, and the kinds of relationships we forge in our practice.

SWAMPY LOWLANDS OF PRACTICE

Helen Simons and Jennifer Greene (2018) envision democratic evaluation like this:

> It can never be a question of whose side the evaluator is on. It can never be a question of colluding with the most powerful of stakeholders. It can never be a question of excluding marginalized stakeholders just because it is logistically difficult to include their experiences and viewpoints. It can never be a question of caving in when an evaluation report is criticized or attempts are made to censor it. There are always multiple perspectives to be heard and represented. And it is indeed a fine-tuned juggling act to maintain faith with all stakeholders while favoring none, to negotiate the inclusion of different values, and to attend to legitimately different interests. Democratic evaluation is a practical political art. Not for the fainthearted. Nor reliant on simply producing a methodologically sound report. Such a report is a clear evaluative responsibility but cannot, in and of itself, advance the values of democracy. (p. 89)

The characterization of evaluation as a "practical political art" speaks to our imagination because it refers to the power of the evaluation practice. In the double sense of the word, evaluation is indeed full of power as well as powerful. Let's explore what this vision means in practice. We start with a practical example because we are convinced that complex, ethical-moral questions arise out of daily practice, not in theory. In practical situations, evaluators as practitioners are confronted with the complexities of their work, the so-called "swampy lowlands" (Schön, 1983, p. 42). The swampy practical situations we have in mind here are indeed sticky and murky and uncomfortable places to be. Professionals longing for control and predictability avoid and stay out of these places. These are typically the situations wherein practitioners cannot rely on handbooks and science-based knowledge. Science-based or instrumental knowledge is about progress, technological innovation, and maximal controllability. Donald Schön (1983) refers to this domain as the "high grounds" (p. 42). High grounds offer stability and certainty but cannot provide "solutions" for the difficult and untamed problems practitioners encounter in practice. Indeed, some of these problems, those of moral nature, may not even be fixed at all. The particulars of the situation require juggling, and another type of knowledge, knowledge of relational and moral nature. Although the swampy lowlands are difficult places, in his book on the reflexive practitioner, Schön (1983) characterizes these lowlands as the places of the "greatest human concern" (p. 42), so they matter to many of us:

> In the varied topography of professional practice, there is a high, hard ground where practitioners can make effective use of research-based theory and technique, and there is a swampy lowland where situations are confusing "messes" incapable of technical solution. The difficulty of the problems of the high ground, however great their technical interest, are often relatively unimportant to clients or to the larger society, while in the swamp are the problems of greatest human concern. (p. 42)

The practical example we draw on throughout this chapter is illustrative of epistemic injustice and appropriate to address the issue of responsibilities. It goes back to 1990 when I, Tineke, recall one of my first evaluation studies, part of my PhD, where I was asked to evaluate a rehabilitation program for psychiatric patients. The management of a psychiatric hospital commissioned the evaluation. The rehabilitation program aimed to focus on the capacities of patients and stimulate them to find a meaningful daily activity outside the hospital. It seemed no more logical to me to include the patients, but when I proposed to include them in the evaluation, the therapists warned me that the people involved were psychiatric patients and that they might become psychotic if I engaged them in an evaluation. Moreover, the therapists showed that they already had measured patients' satisfaction and knew what patients needed. For the first time, I was confronted with the power and external pressure of experts. I realized how this could lead to the exclusion of the voices of those who were served and for whom the program was developed in the first place. This experience is a situation wherein patients are wronged in their capacity as a knower.

Young and idealistic as I was, I felt I had to stand up for my beliefs and persuade the therapists and the commissioners that patients needed to have a say and had valid knowledge grounded in the lifeworld experiences. Based on standard questionnaires, I felt a strong internal drive to resist the therapists in their persuasion that they already knew what patients' needs and desires were. What did these numbers tell us about patients' experiences? The urge I felt was grounded in informal conversations I already had with several patients. In these conversations, they shared some serious concerns with me about the rehabilitation program. For example, they expressed concern over working toward and accepting a regular job in an organization outside the hospital. This was a normative ideal set by the hospital management and was part of the social inclusion policies at the time, but clearly did not fit with most of the patients who were happy with the recreational work done in the protective environment of the hospital. At the time, I was a novice and a bit anxious about the risks the therapists warned me about. Yet, I also felt I could not let the patients' appeal on me go unanswered. So, I started a conversation with the managers and therapists about their views of me as a researcher, the patients, their concerns, and underlying value commitments. As a qualitative evaluator, I explained that I was planning for conversation-like interviews about their daily experiences and intended not to ask intrusive questions about their past (trauma). I also arranged for a therapist to be available after the interview if patients wanted to talk, a suggestion I received from one of the younger psychologists at the hospital with whom I had a clique and built up a friendly relationship. In retrospect, I think the availability of a therapist convinced everyone concerned that what I did was safe and in control. Finally, I got a green light to include the patients in my evaluation.

In retrospect, we recognize the "ethics work" I had performed; I stood for what I valued (social justice and inclusion) and fought in a dignified way for the values that were and are part of my moral horizon. Needless to say, new dilemmas arose during the process. I was lucky that besides the young psychologist, I had a couple of PhD friends with whom I could talk to about these dilemmas. Below I will relate these dilemmas in greater detail, with reflections from both of us.

RECOGNIZING ONE'S RESPONSIBILITY

Democratic evaluation means one has to invest in building trusting relationships, opening spaces for the voices of the marginalized, resisting external pressure, and acknowledging one's own power and non-neutrality.

Gaining Confidence and Building Trustful Relationships

The situation described already gives the reader an insight into the tensions and dynamics in the context. I was hired by the hospital management and had to work hard to become acceptable to all stakeholders if I was willing to engage them. I needed to gain their confidence that I was not just representing the interests and values of the management, but I was also open to the perspectives of other stakeholders. Furthermore, I needed to convince them that I would not judge them and that I was not a traditional academic expert who would tell them what to do. This meant relational work to build confidence among the therapists, the managers, and the patients. Listening to the therapists, I sensed a lot was at stake for them. This was a new program, and the manager had high expectations of the flow of patients to regular workplaces. It also became clear that the time constraints and desired outcomes were set in advance without involving the therapists themselves, but were of great concern to them. The therapists were unsure whether they could achieve the desired outcomes set by the management. There was also a lot at stake for patients, as they could be relocated during the day from the recreational work at the hospital to a regular job. Soon it became clear that some patients experienced the protected workplaces in the hospital as pleasant and had no need to move up. Their first need was to live a good life, contribute in a meaningful way, and have a role.

Opening Spaces for the Voices of People Who Are Less Able to Express Themselves

After I got the green light to include the patients, I planned to interview them to learn about their experiences. However, it soon became clear that the interview hardly provided any interesting information. Patients responded with short cryptic sentences, socially desirable answers, or nothing at all. That silence made me feel uncomfortable. Didn't I ask the right questions? Wasn't I empathetic enough? I reflected on the interview setup, and it struck me that symbolically, this was a situation where I was probably seen as a traditional detached researcher and all-knowing expert who was asking them questions and judging them based on their answers. The setup was hierarchic, and participants resisted and were unwilling to go along with the situation I had created for them. This nonresponse of the patients affected me and prompted me to look for other avenues. After all, I wanted to hear their voices, and in this way, I got no insight into their lifeworld. The situation demanded something different from me; if I wanted to hear their experiences, I had to dare to let go of my interview method. This took some courage. Moreover, I had to deal with patients' expectations and let them experience that I was not a traditional researcher. In other words, my academic frameworks and the images patients had of me as a researcher hindered the inclusion and participation of patients.

In the end, I decided to take a completely different approach. Many of the patients worked in the hospital's landscaping and gardens. They learned to garden there. I thought I could work with them and soon discovered that all kinds of conversations started when I sat down next to them gardening.

For example, I remember one of the guys telling me he thought this was boring work. He found working with devices much more challenging. Most importantly, I got to know them personally: their upbringing, how they ended up in a hospital, how they experienced the hospitalization (most of them were chronic patients), and their needs and dreams. After drawing up a series of personal stories based on this fieldwork, I decided it would be good to organize a focus group with all patients in the rehabilitation program to validate my findings. I wondered how I could do that so that we would not end up in a situation of silence again. The experiences with gardening together gave me the idea to organize a picnic. I tested my idea with some patients I had built a good relationship with, and they confirmed that nature had a calming effect. To give it a special touch, I decided to bring some tasty sandwiches. I knew that not everyone in the group spoke easily and had, therefore, brought some visual material (photographs and paintings of people, animals, flowers, and landscapes) to go beyond the cognitive and verbal. I remember one of the patients choosing a boy on a fence (who he once was) and a large elephant (how he saw himself at the moment). Metaphorically, he had revealed his former and current identity.

In reflection on this experience, we agree with Simons and Greene (2018) when they write, "It can never be a question of excluding marginalized stakeholders just because it is logistically difficult to include their experiences and viewpoints" (p. 89). As the example above shows, there are many more subtle and less subtle barriers to including marginalized groups, in this case, psychiatric patients. Part of these barriers has to do with how we work as academics and the methods and techniques we commonly use to produce knowledge. These methods are not neutral but create a relational distance and verticality between the all-knowing expert/evaluator and the laypeople who are the object of our research in the traditional academic context. In another more recent piece, we have called this the "verbal weaponry" (Bos & Abma, 2018, p. 18) of the academic world . Indeed, much of our methods focus on rational, cognitive, and verbal expressions, assuming that everyone can express themselves through language, and all of our experiences can be captured in words. This cuts off large parts of our consciousness and limits our knowing. Thus, including people in democratic decision-making processes implies that evaluators require sensitivity and a willingness to acknowledge their own responsibility in this process of opening spaces for other kinds of knowing. If we aim to include those who find themselves in a marginalized place, evaluators have to be prepared to search for ways in which they can express themselves, think beyond the usual verbal approaches, and explore new territories, including those that are unfamiliar and uncomfortable, to get into touch with those who are marginalized and their ways of knowing.

So, instead of clinging to the predictability of traditional methods and staying on the so-called high grounds, evaluators need to enter the "swampy lowlands" to figure out what is required to create a space for those whose voices are usually not heard. Tina Cook (2021) aptly describes the "mess" (p. 17) professionals may experience when people they work with give feedback on their methods and how uncomfortable this may be for them. Yet, when people are brought into a situation where they can reflect on their taken-for-granted behaviors and unpick their usual responses, habits, and customs, new territory may open up:

> The value of unpicking accepted knowings was that it provided conditions for shaping new knowings. Accepting "not knowing" could however, leave people feeling in a mess: not knowing what to think or do next. This destabilising activity, instrumental in opening up space for creating new possibilities, was likely to be uncomfortable. (p. 17)

The quote beautifully captures the places we prefer to avoid as researchers and how the standard and protocols offer us security and control.

Resisting External Pressure

In the case study, there was external pressure not to include patients in the evaluation. In this case, the external pressure came from the therapists who thought about patients from their medical and psychologizing frameworks and warned me from those normative beliefs and frameworks. The term "psychotic" clearly came from their medical and psychological outlook and vocabulary, but this categorization was neither neutral nor harmless. I knew that patients often experience these labels as stigmatizing; labels reduce people to their illness and bring the multiple identities back to a negatively charged sick role and identity. Furthermore, the therapists' statements that patients could become psychotic revealed a particular value orientation. Patients were not acknowledged as credible knowers. We have encountered this issue in many other projects; professionals agree that it is important to involve patients until it becomes concrete. Then suddenly, all kinds of difficulties are brought to the fore, such as: patients do not want to participate, will not stick to appointments, will not open their door, or will drop out. In short, they are portrayed as unreliable knowers.

I also sensed in this case that the protection of the safety of the patient and the management of risks was the highest good to strive for in the context of the hospital. I wondered if these were also important values for patients and whether measuring patient satisfaction gave insight into the patients' perceptions. I also wondered whether therapists were truly aware of what was going on with patients, as they said they were. In short, I wanted to resist the pressure to exclude patients from the evaluation. The question was how I could do that so that they would not feel insulted and were willing to continue to participate in the evaluation. I decided to ask naïve questions from a position as a novice and took a genuine interest in their concerns.

In retrospect, Michel Foucault's (2001) concept of parrhesia can be helpful to understand this evaluation context. Parrhesia means "free speech" and is about discours and truth (2001, p. 11). Foucault's research of the meaning of parrhesia in the time of the Classic Greeks entails a linkage between *êthos* and the political. In this period, the modern distinction between beliefs and scientific truth had not yet taken hold. Truth was not, as in current times, connected to objectives and scientific evidence, but it was also related to morality: What is the right thing to do? With this concept of parrhesia, Foucault (2001) raises important questions:

> Who is able to tell the truth? What are the moral, the ethical and the spiritual conditions which entitle someone to present himself as, and to be considered as, truth-teller? About what topics is it important to tell the truth? What are the consequences of telling the truth? What are its anticipated positive effects? What is the relationship between the activity of truth-telling and the exercise of power, or should these activities be completely independent and kept separate? (p. 2001, p. 169)

Parrhesia, therefore, connects questions about knowledge, truth, power, and morality—exactly the themes that play an important role in the process of democratic participation and mutual learning processes we aim for in our work. From this notion, knowledge and truth are not limited to a rational, objective perspective on knowledge, but they are also connected to normativity and ethics. Therefore, other forms of knowledge are part of knowledge production: experiences in the lifeworld, the nonverbal, emotions, the normative, and the ethical.

Parrhesia is a way of bringing an issue of social or epistemic injustice to the table and questioning well-rehearsed knowings, taken-for-granted notions, habits, and routines. Bringing up these issues can be seen as parrhesiastic resistance against dominant relations of power (Foucault, 2011). The parrhesiastes is speaking the truth from an ethical point of view and takes the courage to challenge the dominant power or discourse. In our case example, the risk of the evaluator is that she can be seen as a researcher that is not capable of doing her research properly. At the same time, she feels the urge to do it because something is at stake morally: She wants all involved to have the opportunity to express their experiences and strive for epistemic justice. Parrhesia can be confronting and bring people in "the avoided places" (Woelders, 2019 p. 226), but the intention is not to disrespect, blame, replace, or morally stand above others. It is not strategic or given in by power or status, but rather an open, genuine, and honest way of speaking. It is morally driven because something is at stake. The prime intention is to draw attention to something unjust, like leaving patients out from the evaluation in this context.

Evaluators who use parrhesia know they need others to jointly explore situations and find out what is right and just in a situation. While there are tensions and confrontation, there is a willingness to relate and connect. Therefore, they are sharing their own concerns and listening to the concerns of other people. However, this remains a risky endeavor as one can be seen as a difficult person, which may ultimately lead to exclusion or ignorance by those in positions of power. So, when using parrhesia, one needs to have courage but takes this risk from a moral horizon of truth telling and challenging taken-for-granted power constellations that might not be just.

Acknowledging One's Power and Non-Neutrality

Building trust and remaining acceptable for all stakeholders was difficult, especially after explicitly placing the patients' inclusion on the agenda. Some of the therapists doubted whether I was a neutral and independent evaluator who would consider their interests and perspectives. This was the frame I had encountered when I interviewed the patients the first time. This image of the neutral, objective researcher that others have is something we have to work with. For me, this was a confrontation back then. I knew I was not neutral. I would have never done so much work to include the patients in the evaluation if I had been neutral. I had normative ideals, and these were guiding me in my work. However, this was not in line with how other people expected me to behave (as a neutral expert). I also felt uncertain about it, given my socialization in the traditional canon of research at the university. In our later work, we have connected this to the role of power (Woelders, 2019). As researchers and evaluators, we do not stand outside the process of knowledge production. It may take a lot of energy to position oneself in the context of the evaluation projects and in science. It raises this question: "How do I relate to the scientific, academic system myself, with a strong focus on objectivity and neutrality, while, at the same time, having a moral horizon to contribute to social inclusion?" In addition, one becomes aware of the role of power, often in subtle forms that relate to certain discourses and taken-for-granted norms about science and evaluation.

Striving for social inclusion and valuing the knowledge and contribution of all involved does not come easily or without our involvement and a reflection on our normative position. During our evaluation studies, we became aware that we had to "put an interest" (Woelders, 2019, p. 227). This has a double meaning: From the horizon of social inclusion to strive for good care and sound research, we were interested in the perspectives and experiences of all involved, especially in the experiential knowledge of patients. But we also had to put in interest in bringing these perspectives to the fore and make them explicit within health care organizations and a scientific field. We had to use our power to "put an interest," yet in a way that did not lead to a loss of connection. This was possible by reflecting on our position, the taken-for-granted frameworks, and the support from befriended colleagues. This also sheds new light on creating open spaces in the organizations where the system world and strategic communication are dominant; we learned most by entering the swampy lowlands. This was also a meaningful place because of the human concerns that connected us. The issues raised were not easy to solve, such as how to deal with the desire of some of the patients to stay in the hospital for the rest of their life while the rehabilitation program focused on the empowerment of patients and their social inclusion in society. Yet, the reflection on these issues and their underlying values (freedom of choice, autonomy, protection, care) were helpful to understand each other better and feel connected.

CONCLUSION

Democratic evaluation is a practical political art (Simons & Greene, 2018). In line with Greene's contribution to the field, explicitly brought to the fore in her paper called "Evaluation as Advocacy" (Greene, 1997), we built on the idea that the evaluation practice is not value-free and completed in a vacuum. Nor is it a mere scientific endeavor. Democratic evaluation is a value-laden practice emerging in a power-laden context. The supposed open spaces are power-full spaces (Woelders, 2019). The commitment to epistemic justice and social inclusion means that democratic evaluation includes ethics work (Abma, 2020; Abma et al., 2020; Groot, 2021). The emergence of open spaces requires making room for those spaces inside power constellations while also opening the power spaces within oneself. At the same time, in these power-full spaces, the most valuable lessons can be learned about the moral responsibilities and value commitments we, as democratic evaluators, affirm in our practice.

REFERENCES

Abma, T. A. (2020, September). Ethics work for good participatory action research, *Beleidsonderzoek Online*. https://doi.org/10.5553/BO/221335502020000006001

Abma, T. A., Leyerzapf, H., & Landeweer, E. (2016). Responsive evaluation in the interference zone between system and lifeworld, *American Journal of Evaluation, 38*(4), 1–14. https://doi.org/10.1177/1098214016667211

Abma, T. A., Visse, M., Hanberger, A., Simons, H., Greene, J. C. (2020) Enriching evaluation practice through care ethics. *Evaluation, 26*(2) 131–146. https://doi.org/10.1177/1356389019893402

Bos, G., & Abma, T. (2021) Putting down verbal and cognitive weaponry: The need for 'experimental-relational spaces of encounter' between people with and without severe intellectual disabilities, *Disability & Society, 2021,* 1–25. https://doi.org/10.1080/09687599.2021.1899896

Cook, T. (2021, February) Participatory research: Its meaning and messiness. *Beleidsonderzoek Online.* https://doi.org/10.5553/BO/221335502021000003001

Foucault, M. (2001). *Fearless speech* (J. Pearson, ed.). Semiotext(e).

Foucault, M. (2011). *The government of self and others: Lectures at the Collège de France, 1982–1983.* Picador.

Fricker, M. (2007). *Epistemic injustice: Power and the ethics of knowing.* Oxford University Press.

Gadamer, H.-G. (1960) *Wahrheit und methode.* JCB Mohr.

Greene, J. C. (1997) Evaluation as advocacy, *Evaluation Practice, 18*(1), 25–35. https://doi.org/10.1177/109821409701800103

Greene, J. C. (2006). Evaluation, democracy and social change. In I. F. Shaw, J. C. Greene, & M. M. Mark (Eds.), *The SAGE handbook of evaluation* (pp. 118–140). SAGE Publications.

Greene, J. C. (2010). Serving the public good. *Evaluation and Program Planning, 33*(2), 197–200. https://doi.org/10.1016/j.evalprogplan.2009.07.013

Groot, B. (2021) *Ethics of participatory health research: Insights from a reflective journey* [Unpublished doctoral dissertation]. VU University Amsterdam. https://research.vu.nl/en/publications/ethics-of-participatory-health-research-insights-from-a-reflectiv

Habermas, J. (1987). *The theory of communicative action, volume 2: Lifeworld and system: A critique of functionalist reason.* Beacon Press.

Hanberger, A. (2006). Evaluation of and for democracy. *Evaluation, 12*(1), 17–37. https://doi.org/10.1177/1356389006064194

Schön, D. (1983). *The reflective practitioner: How professionals think in action.* Arena, Basic Books.

Simons H., & Greene, J. (2018) Democratic evaluation and care ethics. In M. Visse & T. Abma (Eds.), *Evaluation for a caring society* (pp. 83–104). Information Age Publishing.

Woelders, S. (2019). *Power-full patient participation—Opening spaces for silenced knowledge.* Dissertation, VU University Amsterdam. https://research.vu.nl/en/publications/power-full-patient-participation-opening-spaces-for-silenced-know

Woelders-Peters, S. M. W., & Abma, T. A. (2017). Looking at participation through the lens of Habermas' theory: Opportunities to bridge the gap between lifeworld and system? In M. Murphy (Ed.), *Habermas and social research: Between theory and method* (pp. 122–137). Routledge.

SECTION II

DIVERSIFYING SOCIAL SCIENCE INQUIRY NATIONALLY AND INTERNATIONALLY THROUGH PEDAGOGY AND MENTORSHIP

Section II emphasizes how Greene's contributions disrupted traditional social science by steadfastly cultivating a diversification of the fields of program evaluation and mixed-methods research nationally and internationally through her pedagogy and mentorship. In Chapter 9, Ayesha Boyce and Lorna Rivera outline the development and field testing of the VEE approach and reflect on how it takes up inclusion, diversity, and equity, as well as how Greene's mentorship continues to inspire their evaluation practice and scholarship as women of color. In Chapter 10, Julian T. Williams and Rafiqah B. Mustafaa present Greene's pedagogical model—learn–practice–reflect—and share how it powerfully impacted their trajectories as evaluators and social scientists. Next, in Chapter 11, Alison Mathie explores the intersections of Greene's work with evaluation and a social justice orientation in the field of international development, highlighting her direct and indirect contributions. This section concludes with Melissa R. Goodnight and Cherie M. Avent. In Chapter 12, these authors explore Greene's influence on the development of program evaluation transnationally, particularly toward evaluation's greater service of diversity and democratic pluralism in public programming and social policy worldwide.

CHAPTER 9

ANSWERING THE CALL

Over Fifteen Years of Disrupting STEM Programming Through Values-Engaged, Educative Evaluation

Ayesha S. Boyce
Arizona State University

Lorna Rivera
Algolia

ABSTRACT

Prominent evaluation scholars often contribute to the field in the form of evaluation theory. They provide thought leadership for how evaluation should be conducted and for what justification. Jennifer Greene and colleagues developed and trained evaluators in an evaluation approach that prescribed the values of equity and diversity and was developed to put the voices of those least well-heard at the forefront (Greene et al., 2006). The values engaged, educative (VEE) approach is the culmination of Dr. Greene's scholarship, mentorship, and theoretical musings. In this book chapter, we outline the evolution of the approach including the centering of values and

Disrupting Program Evaluation and Mixed Methods Research for a More Just Society, pages 123–135
Copyright © 2023 by Information Age Publishing
www.infoagepub.com
123

redefinition of being meaningfully educative. We share examples of how using the VEE approach has facilitated the disruption of traditional evaluations of STEM programs by legitimizing the centering of often ignored stakeholder perspectives, allowing space for stakeholders to reflect on their own bias, and requiring context-specific explicit examinations of inclusion, diversity, and equity. Finally, as women of color, we reflect upon the profound honor of being mentored by Dr. Greene.

Prominent evaluation scholars often contribute to the field in the form of evaluation theory. They provide thought leadership for how evaluation should be conducted and its justification. Dr. Jennifer Greene and colleagues developed and trained evaluators in an evaluation approach that prescribed the values of equity and diversity and was developed to put the voices of those least well-heard at the forefront (Greene et al., 2006). Her vision for this approach was born out of a belief that advocacy within evaluation is inevitable and that evaluators should be explicit about their value commitments and should facilitate stakeholder learning (Greene, 1997).

The development, field-testing, and implementation of the values-engaged, educative (VEE) approach (Boyce, 2017; Greene et al., 2011; Greene et al., 2006; Reid, 2020) has disrupted traditional evaluations of science, technology, engineering, and mathematics (STEM) programs by requiring context-specific explicit examinations of diversity and equity, and has deeply influenced and impacted both of us. Dr. Greene was Boyce's PhD and dissertation advisor. Rivera was introduced to Dr. Greene and the VEE approach while working at the Illinois Science, Technology, Engineering, and Mathematics (I-STEM) Education Initiative at the University of Illinois at Urbana–Champaign (UIUC) with Dr. Lizanne DeStefano as director. As a mentor, colleague, and scholar, Dr. Greene has importantly contributed to our understanding of advocacy and education, attention to culture, diversity, equity, inclusion, and the role of values within evaluation, especially those situated in STEM education and contexts with minoritized populations. We both use the VEE approach in our praxis.

In this book chapter, we will contemplate and outline the evolution of the values engaged, educative (VEE) approach including the centering of values and redefinition of being meaningfully educative. We share examples of how using the VEE approach has facilitated the disruption of traditional evaluations of STEM programs by legitimizing the centering of often ignored stakeholder perspectives, allowing space for stakeholders to reflect on their own bias, and requiring context-specific explicit examinations of inclusion, diversity, and equity. Finally, as women of color we reflect upon the profound honor of being mentored by Dr. Greene.

THE VALUES-ENGAGED, EDUCATIVE APPROACH

Evaluators have been increasingly deliberate about anchoring their work in inclusive, democratic, and emancipatory principles (Greene, 2006); this has been exceptionally true after the murder of George Floyd in May 2020. While not mentioned in all compilations of evaluation theories (Alkin & Vo, 2017), evaluation approaches that encompass the goal of attending to cultural complexity, diversity, justice, equity, and human rights are categorized in Mertens and Wilson's (2018) social justice "branch." Theorists associated with this branch argue program evaluation can embody the values of a more just society and should be positioned as a social, cultural, and political force to address issues of inequity, while still maintaining methodological rigor and trustworthiness (Boyce, 2019). Such evaluation frameworks and theories include cultural and contextually responsive approaches (Frierson et al., 2010; Hopson, 2009; Madison, 1992), transformative participatory (Cousins & Whitmore, 1998), restorative justice (Chouinard & Boyce, 2018), equity-focused (Segone, 2011), transformative (Mertens, 1999), democratic (Kushner, 2005; MacDonald, 1976), deliberative democratic (House & Howe, 2000), and critical evaluation (Everitt, 1996; Fay, 1987). The VEE approach is also situated within this branch.

The VEE approach is the culmination of Dr. Greene's scholarship, collaborations, mentorship, and theoretical musings. As previously stated, her vision for this approach was born out of a belief that advocacy within evaluation is inevitable and that evaluators should be explicit about their value commitments and should facilitate stakeholder learning (Greene, 1997). The VEE approach criteria for high-quality educational programming (especially within STEM) effectively incorporate cutting-edge content; strong instructional pedagogy; and attention to diversity, equity, and cultural issues. Drawing primarily from responsive (Stake, 2003) and democratic (House & Howe, 1999) approaches in evaluation, the VEE evaluation approach calls for purposeful attention to values. The VEE approach advises evaluators to include multiple value stances by encouraging conversation across multiple stakeholder groups. Further, targeted attention to the values of (a) diversity (defined as both traditional socio-demographic markers such as class, gender, and race, as well as other ways people are different from one another) and (b) equity (defined as parity in program access, participation, and accomplishment for all program participants, especially those least well-served in the context) is recommenced by this approach.

Directly influenced by Lee Cronbach's compelling promulgation for evaluators to adopt an educative role (Cronbach et al., 1980), the values engaged, educative evaluator is obliged to facilitate stakeholder learning about the program itself (Greene et al., 2006):

In the VEE approach, evaluators promote critical engagement about the program's theory or underlying logic of activities, outcomes, and interconnections from the primary stakeholders. It is often the case that there are multiple program theories underlying a particular educational intervention. (Greene et al., 2011, p. 24)

Different stakeholders may have differing views on the goals and logic of the program. VEE evaluators are encouraged "to 1) capture diverse stakeholder understandings of the program theory, 2) bring the voices of those less heard to the forefront, and 3) promote dialogue and reflection" (Greene et al., p. 11). All of the previous are examples of efforts to facilitate learning and enhance the educative aspects of this approach. VEE evaluators are also urged to consider using alternative methods, including skits, poems, music, and dance, to present evaluation findings (Boyce et al., 2011; Johnson et al., 2013). The authors of the VEE approach suggest that those who endeavor to use this approach will be experienced evaluators who have credibility, authority, and presence in the evaluation context (Greene et al., 2011). Therefore, the ideal VEE evaluator will have access to program materials, have sufficient opportunities to meet with stakeholders, have expertise in the field (or access to it), and have evaluation team members with similar lived experiences to program participants.

VEE APPROACH EVOLUTION

The VEE approach has evolved since its advent in the 2006 Critical Issues in STEM Evaluation *New Directions for Evaluation* volume (Greene et al., 2006). Much of the changes result from field testing in a variety of STEM contexts and significant reflection by its creators and users. The 30-plus field tests have ranged from small-scale, single-site, limited-budget evaluations to large-scale, multisite evaluations with substantial budgets. Some field test sites have included evaluations of National Science Foundation (NSF, 2016a) funded centers, training initiatives for graduate and undergraduate students, K–12 after-school programming, cyber platforms, and cyber-infrastructure projects. VEE users have urged evaluators to explore alternative approaches for presenting evaluation results (Boyce et al., 2011; Johnson et al., 2013) and embrace prescriptive valuing (Hall et al., 2011). Research on the VEE evaluation approach has investigated how well the approach meaningfully engages the NSF Broadening Participation agenda, offers information of value and relevance regarding connections to underrepresented minorities, and offers information of consequence (Boyce, 2017; Reid, 2020). Empirical research on this approach has also investigated which components or strands of VEE, most importantly, account for successes in various contexts (Boyce, 2017). Next, we briefly discuss the evolution of two components of the VEE approach.

Branding and Centering of Values

The VEE approach debuted with the much catchier name of EVEN, the Educative, Values ENgaged approach. Dr. Greene and colleagues were motivated by Lee Cronbach's compelling promotion for evaluators to take an educative role to aid in stakeholder learning (Cronbach et al., 1980). However, after a few years of reflection, Dr. Greene felt that she needed to rebrand, placing engagement with values, particularly diversity and equity, at the forefront of the approach. Greene debated whether to change the name, as she had already published multiple papers with the approach designated as EVEN. She ultimately felt it was necessary and unveiled the relabeling in the approach guidebook (Greene et al., 2011). Further, as we have developed evaluation competencies, gained experience as evaluators, and grown as scholars, we have infused higher education and social science literature about microaggressions (McCabe, 2009; Sue, 2010), anti-deficit frameworks (Chun & Evans, 2009), and intersectionality (Collins, 2000; Crenshaw, 1989) to inform our thinking around these critical issues and how they impact our practice.

Educative

All evaluators seek access, credibility, trust, respect, and authority (Alkin & Vo, 2017) within their evaluation work. VEE evaluators are called to promote and sustain critical reflection and respectful dialogue to enhance stakeholder program understanding and support improvement efforts. Dr. Greene and colleagues advanced the importance of educative roles evaluators can play. Their focus was on evaluators assisting stakeholders in learning about their programs. We, manuscript authors, have taken the stance that the VEE approach should also encourage stakeholder understanding of evaluation itself, or rather we aim to develop our clients' evaluative thinking (Archibald et al., 2018).

CASE EXAMPLES

In this section, we share examples of how using the VEE approach legitimized the centering of often ignored stakeholder perspectives and allowed space for stakeholders to reflect on their own biases. We also reflect upon our increased ability to disrupt traditional evaluations of STEM programs by requiring context-specific explicit examinations of inclusion, diversity, and equity by employing the VEE approach. All case examples are based on real evaluations conducted by authors.

Case 1: The ACE Staff Climate Study

Program and Evaluation Overview

The Advanced Computing Ecosystem (ACE) was founded in 2011 with funding from the NSF to coordinate the sharing of advanced digital resources like supercomputers and high-end visualization and data analysis resources to enhance researcher productivity. ACE comprises approximately 250 staff employed across 17 institutions within its collaborative partnership. In the project's second year, ACE requested an annual organizational climate study from their external evaluation team to better understand working conditions and staff satisfaction within its unique, large-scale multi-site organizational structure. This request came following the project's second annual NSF Panel Report proactive statement declaring the need for more explicit attention to human-resource-related issues, as they are notoriously problematic for the high-performance computing (HPC) domain. The ACE project management was also particularly concerned with effectively identifying, developing, onboarding, and retaining the appropriate talent to meet the project's long-term goals and objectives.

In response to these requests and goals, we constructed an initial online survey instrument to serve as the foundation for the new Annual Staff Climate Study and establish a baseline assessment of ACE staff's attitudes and concerns within the project so that management could proactively respond and assess change over time. In an effort to protect anonymity and gain trust, the initial study was designed to be anonymous. Due to the lack of diversity in HPC overall and in ACE, there were initial concerns for including gender, race, or ethnicity questions as part of the demographic data collection as they may identify respondents. While we shared these concerns, it also recognized the importance of collecting this data to ensure that traditionally underserved groups were experiencing the project similarly to majority respondents. Given the longitudinal plans for the study and willingness to modify the instrument for the next dissemination year based on the information learned during the Year 1 pilot, we agreed to remove potentially identifying demographic questions from the study.

VEE in Action: Centering Multiple Stakeholder Perspectives

During data analysis, we uncovered qualitative data that suggested issues with gender equity within the organization. However, the evaluation team was limited in our ability to further understand the context given the study's anonymous nature and lack of demographic data.

In response to the findings and to increase the robustness of future climate studies, we set out to present the qualitative data to program leadership as support for designing a more detailed instrument with expanded demographic questions. With the VEE approach as a guide, we acknowledged anonymity concerns during this presentation and devised reporting criteria that would protect respondent confidentiality while also encouraging project leadership to offer suggestions following their own reflection on the findings. This presentation also legitimized the often-silenced voices of communities who may be fearful of sharing their experiences. Lastly, the lack of data from underrepresented communities in this context had led to false understandings of participant experiences that were clarified during this reflection which also encouraged respectful future investigation and response.

Case 2: An Advanced Computing Bootcamp

Program and Evaluation Overview

Since June 2011, federally sponsored programs from North America, Europe, and Japan have collaborated to offer an all-expenses-paid program for graduate students and postdoctoral scholars worldwide who use advanced computing in their studies. Each year, leading computational scientists and HPC technologists from the aforementioned regions offer instruction on various topics over the course of 6 days at the host site, typically a university or research computing center. Bootcamp activities include (a) pre-event attendee preparation through online tutorials and reference materials, (b) tutorials, (c) workshops/hands-on sessions, (d) mentor/mentee program, and (e) social activities.

The evaluation team first began evaluating the bootcamp in 2013 in preparation for its third event. It was conducted on both a formative and summative basis to provide valid and useful information to program leadership, program managers, and domestic and international funders to guide program improvement, assess short- and long-term effectiveness and impact, and increase the likelihood of sustainability. The bootcamp's activities and goal of preparing a larger and more diverse international advanced computing workforce were the focus of the evaluation, which uses a VEE approach.

VEE in Action: Making Space for Stakeholder Reflection

Sustainable funding was a major concern for the bootcamp's leadership and a primary motivator for involving the evaluation team. Given historically high ratings and the novelty of its international component, program coordinators believed their program to be valuable as is and did not see much room for "innovation" as is typically required for continued grant funding from organizations like the NSF. Findings from the evaluation, however, demonstrated significant differences in applicant experience based on gender. Specifically, the evaluation found that applicants self-identifying as women, for example, were consistently rated lower than male applicants.

While the program's pedagogy and scientific content were of high quality, the evaluation team challenged the belief that the program overall was exemplary due to differences in equity and diversity between groups. Using the VEE approach, we had a framework for pursuing investigations into equity, inclusion, and diversity and justified it by highlighting the funding agency's values regarding broadening participation in STEM.

These findings were initially criticized by program staff given they did not relate to the scientific content of the bootcamp, which was valued as the most important, and in some cases, the only determinant of program quality by which the program should be evaluated. As VEE evaluators, we captured the diverse stakeholder understandings of the program theory and responded by including evidence supporting the high-quality scientific content present in bootcamp formal reports. This evidence, however, was paired with student selection outcomes data demonstrating a need for revising the application and selection process, thus bringing the voices of those less heard to the forefront. Lastly, including these findings in the formal report promoted dialogue and reflection around the findings that led to a collective plan devised by the program planning committee with supporting data from the evaluation team to revise their application and selection process. We also presented these data to bootcamp stakeholders before formalizing the final evaluation report to allow time for reflection and discussion around the findings as well as sufficient time to formulate a planned response once the report was publicly released. Lastly, making the report available to the broader community validated and legitimized the often-ignored perspective of underrepresented students.

Case 3: An Evaluation of an HBCU Undergraduate Program

Program and Evaluation Overview

To meet STEM demands and increase underrepresented minorities' degree completion, the NSF provides awards to historically Black colleges and universities (HBCUs) dedicated to enhancing the quality of STEM education and research. The awards are available to "develop, implement, and study evidence-based innovative models and approaches for improving the preparation and success of HBCU undergraduate students so they may pursue STEM graduate programs" (NSF, 2016b, para. 2). An HBCU in the southeastern region of the United States was awarded an HBCU project grant. The short-term goals of the grant were to infuse polymer science into the core content of general and physical chemistry content using a multi-pronged approach to improve study habits, enhance class performance, encourage undergraduates to participate in polymer research, recruit more students to STEM majors, and increase retention and success of students in chemistry. Long-term, the goals are to develop strong ties to soft matter expertise partners in the nearby research park.

The external evaluation team worked with project leadership during the proposal phase until the grant ended 3 years later. The evaluation team used a VEE approach. The VEE approach calls for explicit attention to diversity and equity. As such, the team included the following questions within the evaluation plan: "How and in what ways are project leadership attending to diversity, equity, and cultural issues for participants in activities and throughout the comprehensive program?" and "What opportunities and barriers exist?"

VEE in Action: Context-Specific Explicit Examinations of Diversity and Equity

One of the unique aspects of this context was that all participants and project leaders were all from groups defined as traditionally underrepresented minorities in STEM. While we recognize that the Black experience is not monolithic, the evaluation team initially had some difficulty defining whose voice was least well-served and had to rethink what diversity and equity meant at an HBCU. We spent the first year examining implementation and effectiveness of the project and disaggregated the data by multiple participant identities. We found that project leaders had an unimpeded awareness of their students' different identities, even though they all identified as Black. They also differentiated their mentoring to each student's needs. In addition, we sought to develop a deep understanding of the context itself as none of the evaluators had attended an HBCU.

In Years 2 and 3, we began to examine the extent to which project leadership prepared students to go out into the world where they would once again be underrepresented minorities. Project leaders' educational and industry experiences had heightened their sensitivity to graduating students' needs, particularly around diversity, equity, and cultural issues. While they are located at an HBCU, there was recognition that they are situated within the larger scientific context of preparing and motivating traditionally underrepresented populations in science to seek employment or obtain graduate degrees in STEM. Therefore, they spent a great deal of time informally discussing contexts outside of HBCUs, code-switching, and STEM field etiquette. Project leaders also attempted to challenge students' worldview beyond their current institutional setting in preparation for experiences outside the HBCU context. Had we not employed the VEE approach, we might not have attempted to develop a deep and nuanced understanding of these context-specific but often informal outcomes related to diversity, equity, and cultural issues.

CONCLUSION

Dr. Greene and the VEE approach have changed the very fabric of our professional being. As other authors in this book, notably Williams and Mustafaa, have reflected, Dr. Greene has impacted us tremendously. Further, the use of this approach has facilitated reflection (Dewey, 1933), which has assisted us in unearthing our professional values. Novice evaluators are faced with a plethora of theoretical evaluation approaches to choose from when beginning their formal and informal evaluation training. As evaluators of color, we came to the field with implicit commitments to diversity, equity, and inclusion (DEI; Reid et al., 2020). However, we did not know that we would have the privilege of being trained in an approach that would not only give voice to the least well-served participants, but also allow us as evaluators to be explicit about our values and commitments to being change agents, educative, transparent, and authentic. Over the past 13 years, we have learned that not all evaluators share social justice commitments, and few evaluation training programs offer courses in culturally responsive approaches to evaluation (LaVelle, 2018).

Both of our training in evaluation at the University of Illinois at Urbana–Champaign, Rivera's on-the-job training, and Boyce's PhD training, allowed us to envision a field that challenges the status quo, cares about the voices of those least well-heard, advocates for social justice, and addresses inequities in our work. Further, as Mathie has also argued, Greene's work has encouraged us to be methodological pluralists. The VEE approach has given us a framework for implementing these ideals in our praxis. All of our evaluation projects include questions that address DEI and examine climate with surveys and interviews. As a result, we both feel comfortable discussing social justice, diversity, and inclusion issues with our clients, funders, and other project stakeholders. Evaluators working in STEM fields have long called for attention to culture and DEI (Greene et al., 2006; Mertens & Hopson, 2006) because, as evidence suggests, STEM fields have been riddled with biases (Committee on Equal Opportunities in Science and Engineering, 2017; Lee, 2015). We feel privileged to be among a group trained to and in a position to answer our colleagues' calls.

REFERENCES

Alkin, M. C., & Vo, A. T. (2017). *Evaluation essentials: From a to z.* The Guilford Press.

Archibald, T., Sharrock, G., Buckley, J., & Young, S. (2018). Every practitioner a "knowledge worker": Promoting evaluative thinking to enhance learning and adaptive management in international development. *New Directions for Evaluation, 2018*(158), 73–91. https://doi.org/10.1002/ev.20323

Boyce, A. S. (2017). Lessons learned using a values-engaged approach to attend to culture, diversity, and equity in a STEM program evaluation. *Evaluation and Program Planning, 64,* 33–43. https://doi.org/10.1016/j.evalprogplan .2017.05.018

Boyce, A. S. (2019). A re-imagining of evaluation as social justice: A discussion of the Education Justice Project. *Critical Education, 10*(1), 1–19. https://doi.org/ 10.14288/ce.v10i1.186323

Boyce, A. S., Jimenez, M. B., & Juarez, G. (2011). Presenting cultural and other sensitive evaluation findings through skit. *American Evaluation Association 365 Blog.* https://aea365.org/blog/ayesha-boyce-maria-jimenez-and-gabriela -juarez-on-presenting-cultural-and-other-sensitive-evaluation-findings-through -skits/#comments

Chouinard, J. A., & Boyce, A. S. (2018). Creating collaborative community practices through restorative justice principles in evaluation. In R. Hopson, F. Cram, & R. Millett (Eds.), *Tackling wicked problems in complex evaluation ecologies: The role of evaluation* (pp. 129–154). Stanford University Press.

Chun, E., & Evans, A. (2009). *Bridging the diversity divide: Globalization and reciprocal empowerment in higher education.* Wiley Periodicals.

Collins, P. H. (2000). *Black feminist thought: Knowledge, consciousness, and the politics of empowerment.* Routledge.

Committee on Equal Opportunities in Science and Engineering. (2017). Biennial report to Congress 2017–2018 investing in diverse community voices. Retrieved from https://www.nsf.gov/od/oia/activities/ceose/reports/2017 -2018-ceose-biennial-report-508.pdf

Cousins, J. B., & Whitmore, E. (1998). Framing participatory evaluation. *New Directions for Evaluation, 1998*(80), 5–23. https://doi.org/10.1002/ev.1114

Crenshaw, K. (1989). Demarginalizing the intersection of race and sex: A black feminist critique of antidiscrimination doctrine, feminist theory, and antiracist politics. *University of Chicago Legal Forum, 1989*(1), 139–167.

Cronbach, L. J., Ambron, S. R., Dornbusch, S. M., Hess, R. D., Hornik, R. C., Phillips, D. C., Walker, D. F., & Weiner, S. S. (1980). *Toward reform of program evaluation.* Jossey-Bass.

Dewey, J. (1933). *How we think: A restatement of the relation of reflective thinking to the educative process.* DC Heath & Company.

Everitt, A. (1996). Developing critical evaluation. *Evaluation, 2*(2), 173–188. https:// doi.org/10.1177/135638909600200204

Fay, B. (1987). *Critical social science: Liberation and its limits.* Cornell University Press.

Frierson, H. T., Hood, S., Hughes, G. B., & Thomas, V. (2010). A guide to conducting culturally responsive evaluations. In J. Frechtling (Ed.), *The 2010 user-friendly handbook for project evaluation* (pp. 75–96). National Science Foundation.

Greene, J. C. (1997). Evaluation as advocacy. *Evaluation Practice, 18*(1), 25–35. https://doi.org/10.1177/109821409701800103

Greene, J. C. (2006). Evaluation, democracy, and social change. In I. F. Shaw, J. C. Greene, & M. M. Mark (Eds.), *The SAGE handbook of evaluation* (pp. 118–140). SAGE Publications.

Greene, J. C., Boyce, A. S., & Ahn, J. (2011). *Values-engaged, educative evaluation guidebook.* AEA eLibrary. https://comm.eval.org/thoughtleaders/viewdocument/eval11-session-316

Greene, J. C., DeStefano, L., Burgon, H., & Hall, J. (2006). An educative, values-engaged approach to evaluating STEM educational programs. *New Directions for Evaluation, 2006*(109), 53–71. https://doi.org/10.1002/ev.178

Hall, J. N., Ahn, J., & Greene, J. C. (2011). Values engagement in evaluation: Ideas, illustrations, and implications. *American Journal of Evaluation, 33*(2), 195–207. https://doi.org/10.1177/1098214011422592

Hopson, R. K. (2009). Reclaiming knowledge at the margins culturally responsive evaluation in the current evaluation moment. In K. Ryan & B. Cousins (Eds.), *International Handbook of Educational Evaluation.* SAGE Publications.

House, E. R., & Howe, K. R. (1999). Values in evaluation as social research. SAGE Publications.

House, E. R., & Howe, K. R. (2000). Deliberative, democratic evaluation. In K. E. Ryan & L. DeStefano (Eds.), *Evaluation as a democratic process: Promoting inclusion, dialogue and deliberation.* (Vol. 85, pp. 3–12.) Jossey-Bass.

Johnson, J., Hall, J., Greene, J. C., & Ahn, J. (2013). Exploring alternative approaches for presenting evaluation results. *American Journal of Evaluation, 34*(4), 486–503. https://doi.org/10.1177/1098214013492995

Kushner, S. (2005). Democratic theorizing: From noun to participle. *American Journal of Evaluation, 26*(4), 579–581. https://doi.org/10.1177/1098214005281357

LaVelle, J. M. (2018). *2018 Directory of Evaluator Education Programs in the United States.* University of Minnesota Libraries Publishing. https://conservancy.umn.edu/handle/11299/200790

Lee, A. (2015). An investigation of the linkage between technology-based activities and STEM major selection in 4-year postsecondary institutions in the United States: Multilevel structural equation modelling. *Educational Research and Evaluation, 21*(5–6), 439–465.

MacDonald, B. (1976). Evaluation and the control of education. In D. Tawney (Ed.), *Curriculum evaluation today: Trends and implications* (pp. 125–136). Macmillan.

Madison, A. M. (Ed.). (1992). *Minority issues in program evaluation: New directions for program evaluation.* Jossey Bass.

McCabe, J. (2009). Racial and gender microaggressions on a predominantly-White campus: Experiences of Black, Latina/o and White undergraduates. *Race, Gender & Class, 16*(1–2), 133–151. http://www.jstor.org/stable/41658864

Mertens, D. M. (1999). Inclusive evaluation: Implications of transformative theory for evaluation. *American Journal of Evaluation, 20*(1), 1–14. https://doi.org/10.1177/109821409902000102

Mertens, D. M., & Hopson, R. K. (2006). Advancing evaluation of STEM efforts through attention to diversity and culture. In D. Huffman & F. Lawrenz (Eds.), *Critical issues in STEM evaluation* (pp. 35–51). Jossey-Bass.

Mertens, D. M., & Wilson, A. T. (2018). *Program evaluation theory and practice.* The Guilford Press.

National Science Foundation. (2016a). *Broadening participation in STEM.* https://www.nsf.gov/od/broadeningparticipation/bp.jsp

National Science Foundation. (2016b). Historically Black colleges and universities undergraduate program. https://www.nsf.gov/pubs/2016/nsf16538/nsf16538.htm

Reid, A. M. (2020). Applying an educative approach to engage stakeholder values in evaluations of STEM research and education programmes. *Evaluation Journal of Australasia, 20*(2), 103–108. https://doi.org/10.1177/1035719X20918497

Segone, M. (2011). Evaluation to accelerate progress towards equity, social justice, and human rights. In M. Segone (Ed.), *Evaluation for equitable development results. Evaluation working papers #7* (pp. 2–12). UNICEF.

Stake, R. E. (2003). *Standards-based and responsive evaluation.* SAGE Publications.

Sue, D. W. (Ed.). (2010). *Microaggressions and marginality: Manifestation, dynamics, and impact.* Wiley.

CHAPTER 10

OUR EVALUATION EDUCATION WITH JCG

The Learn–Practice–Reflect Model

Julian T. Williams
John D. and Catherine T. MacArthur Foundation

Rafiqah B. Mustafaa
*Collaborative for Academic, Social,
and Emotional Learning*

ABSTRACT

Throughout her distinguished career as a professor and evaluator, Jennifer C. Greene has taught, mentored, and inspired generations of emerging evaluators. Most of these evaluators were her formal departmental advisees; however, many (including us) were not. As graduate students at the University of Illinois, Urbana–Champaign (UIUC), we benefited from Dr. Greene's theoretical genius, methodological mastery, and hands-on approach to supporting emerging evaluators. As we reflect on our graduate school experience, we identified three pedagogical tools Dr. Greene used in our evaluation education—learning, practicing, and reflecting. It was through this learn–practice–reflect model that

---

Disrupting Program Evaluation and Mixed Methods Research for a More Just Society, pages 137–151
Copyright © 2023 by Information Age Publishing
www.infoagepub.com
137

Dr. Greene guided and supported our development as emerging evaluators. All three components are identified in the evaluator training and education literature as relevant for developing key competencies.

The first component of the model was learning about evaluation in Dr. Greene's evaluation theory and methods courses. Through these courses, we were introduced to theoretical perspectives in evaluation and key tensions on issues such as evaluation purpose and methods. We enhanced our ability to think evaluatively and cultivated mixed-methods skills to develop and execute program evaluations. The second component of the model was to practice evaluation. For 3 years, we had the privilege of serving as research assistants on a 3-year science, engineering, technology, and math (STEM) education evaluation project with Dr. Greene. Through this experience, we practiced evaluation design, data collection and analysis, reporting, and use. Additionally, Dr. Greene served on each of our dissertation committees; in this role, she continued to be an invaluable guide, teacher, and thought partner. The third component of the model was reflection. Dr. Greene introduced us to the American Evaluation Association and the Center for Culturally Responsive Evaluation and Assessment. She not only financially supported us to attend each organization's annual conference, but she encouraged us to submit proposals and present at each conference—both of which we did as graduate students. Participating in these conferences helped us share our work, dialogue, and reflect with other evaluation practitioners and scholars. As a result, we developed professional connections we maintain today.

This learn–practice–reflect model of evaluation education has powerfully impacted our trajectories as evaluators and social scientists. As students in a graduate program outside of Dr. Greene's department, we had limited opportunities to gain evaluation experience; therefore, Dr. Greene provided access to otherwise unavailable opportunities. Collectively, our experiences with Dr. Greene laid the foundation to launch an independent applied research consulting company and occupy roles as evaluators in a foundation (Julian) and an education nonprofit (Rafiqah) where we use the skills we initially developed through working with Dr. Greene, and cultivated through the years. Through this work, we have supported government, foundations, postsecondary institutions, school districts, and community-based organizations.

We know Dr. Greene took the development of emerging evaluators seriously. But we don't know if she realized the powerful model she created for supporting the development of emerging evaluators who were not her formal students. As we reflect on our journeys as evaluators to date, we recognize how critical Dr. Greene's learn–practice–reflect model has been for shaping our approaches to our work. We hope detailing this model will inform other teachers of evaluation as they continue to guide and support the development of emerging evaluators.

In June 2011, we arrived at the University of Illinois, Urbana–Champaign, as first-year PhD students in the educational policy studies program. As graduates of the class of 2011, we chose to go straight to graduate school after college. Rafiqah graduated from Penn State University with degrees in political science and sociology. She entered college to become a lawyer with the goal of changing the systems that created and maintained social inequities. Her plan changed when she was accepted into the Ronald E. McNair post-baccalaureate achievement program, designed to prepare undergraduates for doctoral studies. After spending 2 years learning about and conducting research, she shifted from law school to graduate school, with the new goal of using applied research to address inequities in the U.S. public education system. Julian graduated from Denison University with degrees in education studies and English literature. He entered college to become an after-school program executive director. His goal was to positively impact the lives of African American youth by cultivating a love of learning and service. His plan changed after researching mentoring programs through the Summer Research Opportunities Program (a program designed to inspire and prepare students for doctoral study), interning at two after-school programs, and serving as a community organizer with Citizen Action Illinois. These experiences shifted his goal from serving on a one-to-one basis to impacting groups by improving the effectiveness of programs that serve the African American community.

At the University of Illinois, our graduate program required every student to select a research specialization. We both chose evaluation. It seemed like a natural fit given our aspirations. As we began taking courses in the specialization, we realized how lucky we were to have access to what Hopson and Galloway in this volume call the "Illinois School of Evaluation": Jennifer Greene, Katherine Ryan, Thomas Schwandt, and Robert Stake in educational psychology; Stafford Hood in curriculum and instruction; and Denise Hood in education policy, organization, and leadership. Dr. Greene was our most important teacher of evaluation during our time as graduate students. She taught our introduction to evaluation theory and evaluation methods courses, guided us as we served as her evaluation research assistants, and helped us sharpen our ability to improve our practice through reflection. These experiences are the bedrock of our evaluation education, forming what we call the learn–practice–reflect model. These experiences also align with key activities to support emerging evaluators reflected in the evaluator competency and training literature.

This chapter aims to describe our experience with the learn–practice–reflect model and highlight the key skills and competencies that we developed and sharpened because of the model. First, we introduce the *learn* component and describe how Dr. Greene structured her evaluation theory and methods courses to teach students how to think evaluatively. Scholars have highlighted the important role of learning about evaluation in emerging evaluators' development and orientation to the field (LaVelle & Donaldson, 2010; Stufflebeam, 2001). Next, we describe the *practice* component and detail how engaging in a diverse set of activities as research assistants sharpened our evaluation skills. Evaluation literature documents practical evaluation fieldwork as a critical component of developing competence as an evaluator (Dewey et al., 2008; Dillman, 2013; Galport & Azzam, 2017). We conclude with the *reflect* component by describing the introspective practices we engaged in to document our practice, examine our decisions, clarify our values, and improve our approaches to evaluation. Skillful reflection has been identified as a key competency for evaluators (American Evaluation Association, 2018; Ghere et al., 2006). This chapter details concrete examples of *how* Dr. Greene's values-engaged (Greene, 2012b), democratic (Greene, 2000), and mixed-methods approaches to evaluation (Greene, 2015) powered the learn–practice–reflect model and helped us develop as evaluators. The chapter ends with takeaways that teachers of evaluators may consider as they develop or revise evaluator training programs. Like the other chapters in this section, this chapter demonstrates how Dr. Greene intentionally worked to diversify social science inquiry. In this case, it was by supporting (with her time, energy, and funding) the development of two African American emerging evaluators who were not her formal advisees but who expressed a desire to become evaluators.

LEARN

Our primary experience with the *learn* component of this model was as students in Dr. Greene's Introduction to Evaluation Theory and Introduction to Evaluation Methods courses. Evaluation training programs are essential to the profession of evaluation. These programs prepare evaluation students with foundational knowledge and skills to advance the field of evaluation as scholars and as practitioners (Stufflebeam, 2001). LaVelle and Donaldson (2010) assert that evaluators are made, not born. They argue that extended training periods are necessary to introduce emerging evaluators to evaluation frameworks, ethical guidelines, and standards and to provide space for students to cultivate evaluation-specific skills. This section describes the purpose and structure of the two evaluation-specific courses that we took in graduate school with Dr. Greene. We highlight the key activities that facilitated our learning, the evaluation knowledge we gained, and the evaluation skills we cultivated.

Evaluation Theory

Dr. Greene's Introduction to Evaluation Theory course was designed to help students critically examine diverse approaches to evaluating educational and social programs. Students in the course learned about five major genres of evaluation: evaluation for policy making, accountability, learning, contextual understanding, and democratization. The course focused on understanding each genre's assumptions about knowledge, views of social programs and social change, stances on the role of evaluation in society, location of values in evaluation, and intended use of evaluation findings (Greene, 2013). The primary text *Evaluation Roots: A Wider Perspective of Theorists' Views and Influences* (Alkin, 2013) presented the origins of various evaluation theories through personal narratives and analyses written by key theorists. The course used individual and group assignments to facilitate our learning. We highlight one of each to describe the nature and utility of the assignments.

The final course assignment was to examine a key theoretical issue or genre in practice. Rafiqah focused on evaluation for policy making, arguing that a participatory approach to evaluating government programs is ideal for ensuring the evaluations are inclusive and informed by the most important stakeholders—participants. Julian developed a self-paced program theory construction tool for after-school program leaders. The tool was designed to help program leaders articulate their program's theory of change without assistance from an evaluator. An interest that emerged from learning is that community-based after-school programs rarely have funds for evaluation capacity-building support. This assignment, along with other individual assignments, helped us connect evaluation theory to our interests and deepen our understanding of the relationship between theory and practice.

In addition to the individual assignments, the course also included group debates that greatly facilitated our learning. During the 14-week class, there were five debates, each posing a question related to some aspect of evaluation theory. For example, "Should evaluation primarily serve the public good? What does this mean anyway?" and "Is cultural competence important for effective evaluation? And what is cultural competence?" For each debate, students in opposing groups argued their viewpoints. Students in the audience also participated by posing questions to the debaters. To prepare as a debater or audience member, students read assigned texts for the week and used an analytic guide (a tool introduced by Dr. Greene) to document each theory's key tenets, questions about the theory, and critiques of the theory.

Preparing analytic guides, participating in debates, and observing debates was a powerful experience that brought each evaluation theory to life. This process created a space for us to review and critique each theory from multiple viewpoints on several levels. Together, the course's individual and group activities showed us that there are many ways to approach evaluation, with no approach being better than another. Rather, each evaluator chooses their evaluation approach(es) based on context and ideology, often informed by input from key stakeholders.

Evaluation Practice

Dr. Greene's Introduction to Evaluation Methods course was designed to introduce students to the craft of program evaluation (Greene, 2014). Before each course began, Dr. Greene and her teaching assistant recruited a program and worked with the program's leadership to develop a preliminary evaluation plan. Once the course started, students joined three to five classmates interested in a similar component of the evaluation to form an evaluation "team." Throughout the semester, each team carried out a component of a class-wide evaluation project. The course's primary text was *Evaluation* (Weiss, 1998). Each week the main text was supplemented with other literature, including evaluation case studies and evaluator reflections on their practice.

Students strengthened evaluation skills while completing six major course assignments, each a key section of the final evaluation report: (a) program description; (b) evaluation approach, purpose, audience, questions, expected uses; (c) evaluation design and methods; (d) evaluation instruments; (e) data samples or excerpts; and (f) evaluation results and interpretations. To support students in completing these assignments, Dr. Greene used the evaluation plan as a pedagogical tool for understanding the evaluation's purpose and key considerations, as well as implications for each section of the final report. In addition, during each course session, students engaged with an evaluation case example and articles presenting key learnings about evaluation techniques and evaluator reflections on key decisions made in their work.

The course reading materials, sessions, and assignments turned the evaluation plan and report into key tools for learning evaluation methods. What we learned in each weekly class session was not abstract but practical because we applied the learnings to our component of the overall evaluation. While the scope of the evaluation was small, it was the first time we conducted an evaluation from start to finish, were responsible for making key evaluative decisions, and worked in an evaluation team. When learning about evaluation, there is no substitute for conducting an evaluation. We found that the course's step-by-step evaluation design and implementation process was the foundational experience we needed.

The courses—Introduction to Evaluation Theory and Introduction to Evaluation Methods—were the foundation for our formal evaluation education. In theory, we learned about the work of key evaluation theorists; the values embedded in each theory (Greene, 1997); and how to review, make sense of, and critique each theory. In methods, we learned how to make decisions and ask and answer important questions about evaluation approach, questions, criteria, methods, analysis, and dissemination. In both classes, Dr. Greene debunked the myth of a superior methodological approach in evaluation by arguing that qualitative and quantitative methods both have a place in evaluation and should be used when the project's purpose and questions call for them (Greene, 2012a). We learned the power of mixed methods as a tool for incorporating multiple ways of knowing and multiple value stances (Greene, 2005). We also learned how dialogue can be used as a tool for advancing democratic and inclusive forms of evaluation (Greene,

2001). Together, these courses prepared us for our first experience *practicing* evaluation from start to finish outside the classroom.

PRACTICE

Our primary experience with the *practice* component of this model was through our work as Dr. Greene's evaluation research assistants. As emerging evaluators whose prior formal training in evaluation took place in the classroom, we were ecstatic when Dr. Greene offered the opportunity to work on the 3-year National Science Foundation-funded Widening Implementation and Demonstration of Evidence-Based Reforms (WIDER) initiative. Through WIDER, faculty in science, technology, engineering, and math at UIUC aimed to enhance student engagement, improve student learning, improve recruitment and retention of women students and students from racial and ethnic groups historically underrepresented in STEM fields, develop a new teaching culture for gateway STEM courses, and create a transferable model of institutional change for other institutions. Dr. Greene joined the project as a co-principal investigator. As outlined in this volume by Boyce and Rivera, Dr. Greene's approach to values-engaged, educative evaluation (VEE) has often contrasted more typical approaches to evaluating STEM education initiatives by focusing on context and issues of diversity, equity, and inclusion in evaluation design, data collection, data analysis, and communication. Dr. Greene's commitment to the VEE approach shaped the WIDER project evaluation and our experience as evaluators on the project.

Recognizing the need for evaluation fieldwork, Dr. Greene explicitly wrote evaluation research assistants into the WIDER budget and evaluation plan. Evaluation fieldwork experience is a critical component of evaluator training and competency development (Dewey et al., 2008; Galport & Azzam, 2017). Dillman (2013) reported that members of the American Evaluation Association's Graduate Student and New Evaluator topical interest group ranked evaluation fieldwork as the top experience contributing to their skill development. We agree. At the end of WIDER, we viewed ourselves as "real" evaluators, confident that we could design and conduct an evaluation on our own. This section describes the activities Dr. Greene used to support our skill development in evaluation design, data collection, data analysis, and communicating findings.

Evaluation Design

At the project's outset, Dr. Greene drafted an evaluation plan that outlined the evaluation's purpose, guiding questions, quality criteria, methods, and use plan (Mestre et al., 2013). The plan stated that evaluation research assistants would engage in all parts of the formative and summative mixed-methods evaluation (Greene, 2005), serving as semi-internal evaluators participating in weekly meetings with faculty teams implementing the course reforms. As evaluation research assistants, we revised evaluation questions, developed a program theory, and developed data collection instruments.

Data Collection

While we came to the WIDER evaluation with skills in qualitative and quantitative data collection, we had only used those skills in research and in our Introduction to Evaluation Methods course project. Therefore, with WIDER, we applied skills in new ways and honed our competencies as evaluators. First, we interviewed faculty to understand their pedagogy and thoughts on creating a sense of belonging in the classroom. We observed faculty team meetings to understand how they collaborated, how they solved problems, and where they needed assistance. We surveyed and conducted group interviews with students to understand their experiences in the classroom and their thoughts on entering STEM fields. Finally, we conducted individual interviews with women students and students from racial and ethnic groups historically underrepresented in STEM fields to understand their unique experiences in the courses that were a part of the WIDER project.

Data Analysis

We learned new collaborative data analysis techniques while working on the project. For example, Dr. Greene taught us a team process for systematically coding and developing themes from qualitative data. Using interview transcript data, each research assistant coded transcripts based on initial codes, then passed the transcripts to another team member to repeat the process. After that, the evaluation team discussed the initial round of coding and revised the protocol based on initial learnings that emerged from interview data analysis. The evaluation research assistants then repeated the double coding process using the revised codes. At the time, this collaborative data analysis process was new to us, but it was our go-to approach by the end of the 3-year project. Today, it has become a regular part of our qualitative data analysis process, which we've refined over the years while conducting evaluations in various contexts together.

Communication

We developed skills in communicating evaluation findings by writing evaluation reports and presenting the findings to WIDER faculty teams. During the 3-year project, we produced formative reports each semester to share learnings in real-time and summative reports at the end of each year. In addition to these written reports, each evaluation research assistant presented evaluation findings to their assigned faculty teams. Dr. Greene helped us outline the reports, gave feedback on report drafts, and presented findings to the project's leadership to model a way to present evaluation findings. This scaffolded approach helped model different approaches to presenting evaluation findings and building our confidence in our communication abilities.

Our experience *practicing* evaluation as Dr. Greene's evaluation research assistants helped us develop important skills as emerging evaluators. First, the process allowed us to practice evaluation outside of a classroom environment. Second, this experience was our first time being paid as evaluators. This element heightened the stakes because stakeholders expected us to produce useful data and findings novice/graduate student our graduate student status. Third, we had the opportunity to make decisions about the evaluation plan, data collection, data analysis, and communication methods. This helped us develop autonomy as evaluators, tasked with making decisions that shaped our work and the overall evaluation process. Together, these experiences not only helped us develop skills and shaped our competency as evaluators, but they also helped us develop confidence as new evaluators entering the field.

REFLECT

Our experience with the *reflect* component of this model came through three key activities—participation in weekly evaluation team meetings for the WIDER project, use of analytic memos throughout the WIDER evaluation, and participation in annual conferences hosted by the American Evaluation Association. Reflection provides the opportunity to think critically about one's work, ensure that one's practices are aligned to the task at hand, and improve practice in an ongoing manner (Lenburg, 2000; Roche, 2011). For evaluators, incorporating reflection into one's practice has been identified as a key competency (American Evaluation Association, 2018; Ghere et al., 2006). The three experiences we describe in this section were integral to developing reflection as a part of our evaluation practice.

Evaluation Team Meetings

At our weekly WIDER evaluation team meetings, we planned for the work ahead, solved problems, and made decisions. Dr. Greene often drafted the meeting agendas in advance, identifying key items for discussion and decision-making. At the beginning of each team meeting, team members reviewed the agenda to suggest revisions as needed, given recent events or contextual changes. Additionally, we established a rotation where each evaluation team member was regularly responsible for taking meeting notes.

At the beginning of the WIDER project, these meetings focused heavily on logistical matters such as submitting an amendment for the project's approved application with UIUC's institutional review board, adding faculty team meeting days and times to our calendars, and revising the evaluation's guiding questions. As the project progressed, the focus of these meetings shifted to team members providing updates on the progress of their faculty teams and their respective work tasks. For example, when two team members were developing a student questionnaire, a few meetings were spent with other team members providing critical feedback to improve the instrument. As the project matured, team meetings were spent problem-solving. For example, a few evaluation team members noticed that the work of quite a few departmental teams lacked attention to a key project objective—improving the recruitment and retention of women students and students from racial and ethnic groups underrepresented in the participating STEM fields. Evaluation team meetings provided the space for us to discuss this pattern and strategize on how to surface our commitment to this objective in our work as evaluators, approaching "evaluation as advocacy" (Greene, 1997). As a result of this collective problem-solving, our evaluation team ultimately decided to add a component to the evaluation data collection that elevated the voices of students who were women and from racial and ethnic groups underrepresented in STEM fields. We conducted individual interviews and classroom observations as we endeavored to "visit and listen well," as described by Veronica Thomas in this volume, to thoughtfully engage students participating in WIDER classes (key evaluation stakeholders who would ultimately be greatly affected by decisions made by faculty members engaging with the evaluation findings).

Participating in these regular evaluation team meetings taught us an approach for leading an evaluation team. We learned how to create an evaluation project team agenda that moved the work forward while incorporating reflection into the process. We learned how to create a team environment where there was shared responsibility for the work and opportunities for autonomy and ownership. We learned to create a team culture where discussion was substantive, disagreement was accepted, and diverse perspectives were encouraged.

Analytic Memo/Log Tool

The analytic memo/log tool was another tool Dr. Greene used to facilitate reflection throughout the evaluation process (Greene, 1988). The analytic memo/log tool can document key decisions about data collection, data analysis, client communication, and evaluation use in real time. It can also document things the evaluator interprets as interesting or relevant to the overall evaluation.

For example, after observing courses taught by faculty participating in WIDER, we identified instances when the course reforms were not being used as intended, which was relevant feedback for the faculty planning teams. The analytic memo reflection helped us identify and address this issue in future data collection and communications with the faculty stakeholders. As evaluators, we collect numerous sources of data from stakeholders throughout the evaluation process. Without being intentional, an evaluator may not document their thinking throughout the process. The analytic memo/log tool builds reflection into the evaluation process.

American Evaluation Association Conferences

Presenting at annual meetings of the American Evaluation Association was a built-in reflection mechanism during our WIDER evaluation experience. Throughout the WIDER project, Dr. Greene used funds from the project to pay for our AEA membership, registration, travel, and lodging. She also encouraged us to submit a proposal to present every year of the project—and we did. Each year, our WIDER evaluation team identified a possible presentation topic based on a relevant issue that arose from our work. In 2014, we presented on the challenge of being an external internal evaluator. We discussed our team's unique position within the evaluation, key challenges we faced because of our positioning, and our key strategies for "skating on the inside edge" (Greene et al., 2014). In 2015, we presented our theory-practice dilemma. We discussed how the project was deviating from its initial program theory, our assessment of our options for proceeding, and our final decision (Greene et al., 2015). In 2016, we presented on our use of mini-case studies to examine the experiences of students from demographic groups underrepresented in STEM. The presentation highlighted what we learned and how the project's leadership responded (Perez et al., 2016). Participating in the conference provided a space for our team to engage in ongoing reflection in preparation for the session and engage in discussion and thought partnership with other evaluators. Without attending the conference, we would have missed out on the useful insights of the other evaluators in the room. Today, we still value participation in professional conferences as a key way to reflect and engage in dialogue with other evaluators.

Engaging in these reflective activities sharpened our evaluation practice in several ways. First, these activities helped us to normalize the practice of questioning and challenging our thinking and practice. Second, these activities provided us with a model for reflecting as an evaluation team internally and externally. Third, these activities helped us to make connections between our work as evaluators and the broader field of evaluation. By thinking about our practice and navigating various dilemmas, we could draw from literature, engage other evaluators, and offer our learnings to the field. Today, we maintain these practices as key to our work as evaluators and agree that reflection is a necessary part of good evaluation practice.

CONCLUSION

We don't recall discussing with Dr. Greene an explicit process she used to support emerging evaluators. Nor do we recall her using the phrase learn–practice–reflect model to describe our experiences with her. In this chapter, we have attempted to describe our experiences with Dr. Greene and organize them into a model for emerging evaluator education. We close by describing why we believe the learn–practice–reflect model was effective for our development as emerging evaluators and what we think are some critical elements that teachers of evaluators may consider incorporating as they develop or revise evaluator training programs.

First, the model was built on skills we were developing as doctoral students in educational policy studies. Our graduate methodological training included education research design, theory of qualitative methods, case studies, oral histories, questionnaire design, and statistics. These courses improved our ability to systematically ask and answer research questions using qualitative, quantitative, and mixed-methods approaches. These skills allowed us to fully participate in the learn–practice–reflect process.

Second, the model had teacher continuity. Dr. Greene was involved in each step of the process. This continuity deepened our relationship with her and provided the space for us to ask questions and let her know when we needed more support or information. The continuity also helped Dr. Greene to collect data as a teacher. Over time, she was able to understand our interests, strengths, and growth areas and identify where we needed to be challenged and pushed, as well as positioned to shine.

Third, the model was well sequenced. We started in the classroom, moved to fieldwork, and ended in the field of evaluation. This sequence was effective because each experience built on the previous one. Coursework taught us how to think like an evaluator—asking thoughtful questions and making important decisions. Fieldwork provided an opportunity for us to implement what we learned in the classroom, make decisions, take initiative, and have more ownership in the evaluation process. The reflective activities improved our evaluation thinking skills, giving us tools to reflect individually, giving us space to reflect as an evaluation team, and creating opportunities to engage with the broader evaluation community.

Fourth is a feature unlike the first three because it is difficult to replicate—Dr. Greene. When we enrolled in evaluation theory in the Fall of 2013, Dr. Greene was entering her 34th year as a professor (College of Education, 2022). With more than 30 years of teaching evaluation, it's safe to say we entered the class of an experienced evaluation teacher. She has conducted numerous evaluations in diverse contexts throughout her career. She is a well-respected scholar of evaluation who has published widely on evaluation theory and methods, and when we first took a course with her in 2013, Dr. Greene had recently served as the president of the American Evaluation Association in 2011. As second-year graduate students, we knew none of this. As we look back, this fourth element—the Greene element—made a world of difference. As highlighted by Hall and Copple in the introductory chapter and elaborated by authors throughout this volume, Jennifer Greene has had a profound effect on the development of countless evaluators.

Many may read this and think that it is impossible to replicate Dr. Greene. It is. She's one of a kind. But it is possible to replicate the learn–practice–reflect model of evaluator development that we have described in this chapter. We hope the chapter makes this evaluator-development process more explicit and that it may be useful to educators of emerging evaluators and for emerging evaluators who are thinking about their own development.

Dr. Greene, thank you for this amazing opportunity, for being gracious with your time, and for being dedicated to student development. We appreciate you and we celebrate you.

REFERENCES

Alkin, M. C. (Ed.). (2013). *Evaluation roots: A wider perspective of theorists' views and influences* (2nd ed.). SAGE Publications.

American Evaluation Association. (2018, April 5). *AEA evaluator competencies.* https://www.eval.org/About/Competencies-Standards/AEA-Evaluator-Competencies

College of Education. (2022). *Jennifer Greene.* University of Illinois, Urbana-Champaign. https://education.illinois.edu/faculty/jennifer-greene

Dewey, J. D., Montrosse, B. E., Schröter, D. C., Sullins, C. D., & Mattox, J. R. (2008). Evaluator competencies: What's taught versus what's sought. *American Journal of Evaluation, 29*(3), 268–287. https://doi.org/10.1177/1098214008321152

Dillman, L. M. (2013). Evaluator skill acquisition: Linking educational experiences to competencies. *American Journal of Evaluation, 34*(2), 270–285. https://doi.org/10.1177/1098214012464512

Galport, N., & Azzam, T. (2017). Evaluator training needs and competencies: A gap analysis. *American Journal of Evaluation, 38*(1), 80–100. https://doi .org/10.1177/1098214016643183

Ghere, G., King, J. A., Stevahn, L., & Minnema, J. (2006). A professional development unit for reflecting on program evaluator competencies. *American Journal of Evaluation, 27*(1), 108–123. https://doi.org/10.1177/1098214005284974

Greene, J. C. (1997). Evaluation as advocacy. *Evaluation Practice, 18*(1), 25–35. https://doi.org/10.1016/S0886-1633(97)90005-2

Greene, J. C. (2000). Challenges in practicing deliberative democratic evaluation. *New Directions for Evaluation, 2000*(85), 13–26. https://doi.org/10.1002/ev.1158

Greene, J. C. (2001). Dialogue in evaluation: A relational perspective. *Evaluation, 7*(2), 181–187. https://doi.org/10.1177/135638900100700203

Greene, J. C. (2005). The generative potential of mixed methods inquiry. *International Journal of Research & Method in Education, 28*(2), 207–211. https://doi. org/10.1080/01406720500256293

Greene, J. C. (2012a). Engaging critical issues in social inquiry by mixing methods. *American Behavioral Scientist, 56*(6), 755–773. https://doi.org/10.1177/ 0002764211433794

Greene, J. C. (2012b). Values-engaged evaluation. In M. Sergone (Ed.), *Evaluation for equitable development results* (pp. 192–2017). UNICEF.

Greene, J. C. (2013). *Introduction to evaluation theory* [Syllabus]. Department of Educational Psychology, University of Illinois.

Greene, J. C. (2014). *Introduction to evaluation methods* [Syllabus]. Department of Educational Psychology, University of Illinois.

Greene, J. C. (2015). The emergence of mixing methods in the field of evaluation. *Qualitative Health Research, 25*(6), 746–750. https://doi.org/10.1177/ 1049732315576499

Greene, J., Mustafaa, R., Williams, J., Garcia, G., & Gates, E. (2014). *Skating on the inside edge: Being an external internal evaluator* [Conference presentation]. Evaluation 2014, American Evaluation Association, Denver, CO.

Greene, J., Williams, J., Mustafaa, R., & Gates, E. (2015, November 12). *Whither evaluation theory? Challenges from STEM education evaluation practice* [Conference presentation]. Evaluation 2015, American Evaluation Association, Chicago, IL.

Greene, J. G. (1988). Stakeholder participation and utilization in program evaluation. *Evaluation Review, 12*(2), 91–116. https://doi.org/10.1177/0193841 x8801200201

LaVelle, J., & Donaldson, S. (2010). University-Based Evaluation Training Programs in the United States 1980–2008: An Empirical Examination. *American Journal of Evaluation, 31*(1), 9–23. https://doi.org/10.1177/1098214009356022

Lenburg, C. B. (2000). Promoting competence through critical self-reflection and portfolio development: The inside evaluator and the outside context. *Tennessee Nurse, 63*(3), 11–20.

Mestre, J. P., Greene, J., Herman, G. L., Tomkin, J., & West, M. (2013). *Proposal to National Science Foundation for Widening Implementation & Demonstration of Evidence Based Reforms grant* [Grant proposal]. University of Illinois, Urbana–Champaign.

National Science Foundation. (n.d.). *Widening implementation & demonstration of evidence based reforms (WIDER)*. https://www.nsf.gov/funding/pgm_summ .jsp?pims_id=504889

Perez, M., Mustafaa, R., Williams, J., & Greene, J. (2016). *Navigating challenges to using "mini-case study" to examine the experiences of students from groups underrepresented in STEM* [Conference presentation]. Evaluation 2016, American Evaluation Association, Atlanta, GA.

Roche, M. (2011). Creating a dialogical and critical classroom: Reflection and action to improve practice. *Educational Action Research, 19*(3), 327–343. https:// doi.org/10.1080/09650792.2011.600607

Stufflebeam, D. L. (2001). Interdisciplinary Ph.D. programming in evaluation. *American Journal of Evaluation, 22*(3), 445–455. https://doi.org/10.1177/ 109821400102200323

Weiss, C. H. (1998). *Evaluation* (2nd ed.). Prentice Hall.

CHAPTER 11

CROSSING THE THRESHOLD

Jennifer C. Greene's Contribution to Disrupting Evaluation in the International Development Field

Alison Mathie
Coady International Institute

ABSTRACT

Using the idea of liminal spaces or thresholds, I explore how Jennifer Greene has helped people to navigate the boundaries in social science inquiry between the epistemological certainties of post-positivism on the one hand and epistemological pluralism and dialogue on the other. Through her contributions, she has ushered in disruptive new ideas to evaluation, which now have traction in accepted practice, including in the international development field. I outline her contributions and overlay these on the trends in international development evaluation, demonstrating the relevance of her insistence on methodological and paradigm diversification, as well as democratic engagement, in the pursuit of an evaluation practice of consequence for social justice.

Disrupting Program Evaluation and Mixed Methods Research for a More Just Society, pages 153–164
Copyright © 2023 by Information Age Publishing
www.infoagepub.com
153

In disrupting and diversifying social science inquiry, Jennifer Greene takes us into liminal spaces where boundaries are breached and transition to the new is navigated. The idea of liminality is often described in terms of thresholds—the physical spaces between the outside and the inside or the place left behind and the new destination. It is not a comfortable space; it can be ambiguous and disorienting. Looking back on the contribution of Jennifer Greene to the evaluation field in general and to my own work in international development in particular, I am aware that the idea of liminality also seems an apt description of that threshold she helped people to cross between the epistemological certainties of post-positivism on the one side and pluralistic experimentation and dialogue in evaluation on the other. In this chapter, I dig deeper into this contribution, her ushering in disruptive new ideas to evaluation, and reflect on the extent to which these ideas have now transitioned into accepted practice in the international development field.

As a graduate student of Jennifer's in the early 1990s, like many of my cohort, I was entering a PhD program after many years as a practitioner, often impatient with the pace of academia. As compelling as this new opportunity was, there was also a sense of loss. I was also entering a liminal space, dawdling on this new threshold.

Luckily for me, Jennifer was opening a space in the evaluation field to explore. She had just published the first of several papers constituting landmark work on mixed-method evaluation for different purposes (Greene & Caracelli, 1997; Greene et al., 1989); she and her coauthors had identified the possibilities for not only mixing quantitative and qualitative methods but also the paradigms that supported them. Standards for methodological rigor became contested terrain. She challenged the methodological one-upmanship in the qualitative versus quantitative debates and insisted that the preoccupation with method be matched by the *imperative* of a more democratized evaluation practice. In the process, she questioned not only the hegemony of post-positivism in evaluation inquiry but also the practice of evaluation as accountability "upwards." Over time, her sense of urgency for evaluation practice to advance social change (Greene, 2015) added to the weight of her disruptive intellectual contribution. Unwittingly, unintentionally, I had landed in the right place.

Since graduation, the challenge for me professionally has been to push the boundaries of status quo expectations of evaluation practice, especially in the international development field, a field so broad that few can claim to have grasped more than a few slivers of it. There is no question that my experiences have been frustrating. Thankfully, over time, similar efforts by international development academics and practitioners have grown into a substantial critique of, and resistance to, conventional evaluation inquiry and the audit culture that spawns it, most notably by collaborators in the UK, who coined "The Big Push Back" and, more constructively, "The Big Push Forward" (Eyben et al., 2015). In reviewing these critiques, I am struck by the echoes of Jennifer's contributions as part of the global spread of ideas across continents, academic disciplines, and fields of practice experiencing similar frustrations with evaluation.

In the following sections, I first outline the contributions to the evaluation field that Jennifer is best known for. These are then set against trends in evaluation in the international development sector. These discussions lead to an exploration of the intersections between the two, showing both her direct and indirect contributions. Finally, the chapter will show how her contributions can add more value to the critical new work in international development evaluation, pushing it from a liminal space of disruption to broader acceptance.

JENNIFER GREENE'S CONTRIBUTIONS

Building on their earlier work conceptualizing mixed-method designs for different purposes (Greene et al., 1989), Greene and Caracelli (1997) introduced their edited volume *Advances in Mixed-Method Evaluation* by stating:

> Previous mixed-method work has concentrated on the technical level of method, focusing on combining qualitative and quantitative methods within one evaluation study. The present mixed-method volume addresses the philosophical level of the paradigm, analyzing the challenges of combining in one study different, even conflicting assumptions about the nature of social phenomena and our claims to know them. (p. 1)

They noted that there was little argument against the value of employing multiple methods to counteract bias and extend inquiry into corners where one method cannot reach. However, there was heated debate about whether mixing methods attached to different inquiry paradigms could ever be justified. Some, such as Elon Guba and Yvonna Lincoln working in a constructivist paradigm, argued the purist position that paradigms could not be mixed. Others took a pragmatic stance, arguing that whatever worked best to suit the inquiry was justified—the evaluation design should not be inhibited by such philosophical considerations (e.g., Michael Quinn Patton). A third position promoted by the authors was a dialectical one: Paradigmatic differences were important and needed to be explicit "anchors" of knowledge, but mixing methods associated with both could enhance a study by introducing an opportunity for tension, from which greater discovery and mutual understanding could ensue (Greene & Caracelli, 1997,

p. 11). For example, combining interpretivist and post-positivist inquiry traditions could, they argued, yield both emic and etic perspectives, particular and generalizable characteristics, and "social constructions and physical traces," and the resulting knowledge claims were likely to be "more relevant and useful, and more dialectically insightful and generative, even if accompanied by unresolved tensions" (Greene & Caracelli, 1997, p. 13).

This commitment to the dialogue that mixed methods and mixed paradigms could generate became a repeated theme in Jennifer's work and an important signifier of the value she attached to different ways of knowing and different holders of knowledge and their standpoints. As she argued:

> Contemporary social inquiry is challenged by such issues as complexity, contextuality, values and societal role. Rather than debate which stances on these issues are most defensible, a mixed methods way of thinking legitimizes and respects multiple stances and encourages dialogue among them, in service of better understanding. And good social inquiry becomes dialogic, pluralistic and consequential. (Greene, 2012, p. 770)

To be consequential, Jennifer has challenged us as evaluators on two fronts. First, she argues for us to be engaged as self-aware and value-driven inquirers with a commitment to social change. Our task is to prescribe as well as describe values and to be explicit about the values we espouse (Greene, 2011; Hall et al., 2011). This means not only surfacing multiple perspectives through mixing methods but also pushing for social change through the practice of evaluation, democratizing the opportunities for different voices to be heard, and fostering equitable relations within the evaluation exercise itself. In this way, the practice of evaluation takes inquiry into a moral and political space that is necessary rather than avoided as a contamination risk. If intended so, the evaluation process is itself a pathway toward transformation.

Second, to be consequential, evaluation findings (and the evidence on which they are based) have to be convincing. Here, Jennifer expresses concerns about the failure to bridge the gap between mixed-method theory and its application in mixed-method practice. Despite an "explosion" in mixed-methods practice,

> Many empirical researchers and evaluators who use a mixed-method approach do so with little cognizance of the depth of mixed-method theory and thus the potential of this methodology to contribute to credible evidence about consequential solutions to the world's pressing problems. (Greene, 2013, p. 118)

To bridge this gap, Jennifer argues for greater rigor, not just in the technical sense but also in communicating the empirical, political, and social "faces" of the inquiry. The inquirer has to be explicit about the value as-sumptions they bring ("values of distance, engagement, inclusion, objec-tivity, generalizability, contextuality, social action, and so forth" [Greene, 2013, p. 110]) and must also provide a clear rationale for *what* is being mixed in a mixed-methods study and *why*. In the controversial terrain of what constitutes "evidence," she goes further, wanting to

> reclaim the concept of credible evidence from its narrow definition as causal claims regarding intended outcomes and its pristine position as requiring only a methodological warrant, and to reassert the key importance of demo-cratic values in assessing the credibility of evidence and thus also to reframe this assessment as an inclusive, dialogic process rather than a matter of meth-odological purity. Well beyond good method, making meaningful and con-sequential judgments about the quality and effectiveness of social and edu-cational programs requires engagement, interaction, listening, and caring. (Greene, 2015, pp. 217–218)

For Jennifer, taking the potential of mixed methods and paradigms across the threshold into established practice means holding evaluators account-able—not only for methodological appropriateness and rigor but also for cultivating the trustworthy relationships to build "confidence in the merit and value of the evidence generated by the evaluation" (Greene, 2015, p. 209) and for ensuring relevance to all stakeholders in a given context, es-pecially those for whom the program is designed to serve (and, within this group, those potentially least well served). Generated in these ways, cred-ible evidence can render judgments that are "restoratively consequential and action oriented" (Greene, 2015, p. 210).

THE INTERNATIONAL DEVELOPMENT CONTEXT: TRENDS IN EVALUATION

Jennifer's concerns about the unrealized potential of mixed-methods evalu-ation and the narrow definitions of credible evidence resonate with the struggles experienced in the international development sector. Rumblings over many years found a substantial voice in an international conference held at the Institute of Development Studies in the UK, resulting in *The Politics of Evidence and Results in International Development: Playing the Game to Change the Rules?* (Eyben et al., 2015). In this collection, Eyben (2015), for example, provides the background to a "results and evidence agenda" that became particularly influential among donor agencies in the early 2000s, preceded by the use of logic models such as logical framework analysis for planning and management purposes in the United States and elsewhere. Over time, results reporting became more stringent all through the "aid chain" (as it was then described, tellingly), whether a government agency, an international nongovernmental organization (NGO), a local NGO, or a community-based organization (CBO). For some development workers, the shift could be positive as it encouraged well thought out and testable theories of how change happens in any given context that could be revised on an ongoing basis as the situation evolved on the ground. However, such flexibility was not a given (Whitty, 2015). Problematically, as Eyben argues, the "artefacts" of this agenda (such as randomised control trials, social return on investment and impact evaluation, even payment by results) became tools for discipline and control. Planning frameworks limited to linear causality and objectively verifiable indicators to measure evidence of change became straightjackets within which evaluation for accountability was prioritized over evaluation for learning, and projects emphasizing measurable outcomes began to drive out those with immeasurable ones.

In the postcolonial era, this reproduction of relationships of control in the name of value for money and efficiency was deeply troubling. Stories abound of recipients of government aid having to distort and reduce complex programs to the bare bones of measurable indicators to secure funding, with huge opportunity costs in terms of time and energy expended satisfying donor requirements as well as the psychological effects of what could be a harrowing experience (Chambers, 2017). I can add my own stories of trying to help a women's organization come up with measurable indicators of empowerment for an external donor when the very process of compliance was disempowering in itself, or of attempting to identify predetermined indicators of change for community groups self-mobilizing in highly unpredictable environments. Life was unfolding but not necessarily as intended or at the pace and convenience of arbitrary outcome and impact assessment time frames. Similarly, when teaching an evaluation course for development practitioners from the Global South, I was dismayed to discover that many participants were not interested in mixed-method or participatory evaluation so much as how to satisfy donor requirements for a logical framework and measurable outcome and impact indicators. In terms of the metaphor in the title of Eyben et al.'s book, they were learning how to play the game for their own survival; learning how to change the rules of the game seemed more like an intellectual indulgence.

The conundrum for international development evaluation is familiar to the evaluation field more broadly—balancing accountability and learning and trying to merge the two in evaluation research that can reasonably claim to shine the light on causal inference (Picciotto, 2014) while remaining open to the unexpected. Large sums of public funds are disbursed in bilateral aid budgets, with substantial funds also contributed by private charitable foundations and private sector corporations through their corporate social responsibility mandates. One can reasonably claim that it is in all stakeholders' interests to see if the funds are well spent, build a better understanding of what works, and set achievable goals. At a global level, since 2000, expectations for specific results have been set as achievable targets in the United Nations Millennium Development Goals (MDGs) for countries in the Global South, replaced in 2015 with targets for the Sustainable Development Goals (for all UN signatory countries) to be achieved by 2030. Significantly, alongside the MDGs, and to increase the likelihood of success, 100 donors signed on to the Paris Declaration on Aid Effectiveness in 2005, setting targets for an architecture of effective aid delivery for 2010. Included in its mutually reinforcing principles was management by results, with all countries expected to develop assessment frameworks to measure impact[1] (Organization for Economic Cooperation and Development, 2019). This global commitment to management by results helps to explain why it has percolated throughout the development sector.

In agreement with Eyben's observations, several writers (e.g., Gujit, 2015; Mathie & Peters, 2013; Chambers, 2017) claim that the pervasiveness of results-based thinking, however laudable in original intention, has led to a locked-in methodological and paradigmatic bias, with quantitative methods and experimental designs such as RCTs pushed in the service of "the results and evidence agenda" to the detriment of discovery through a more pluralistic methodological approach. Chambers' (2017), for example, shows how the quest for the measurable leads to bias towards a specific finding. This can be a blind spot with potentially serious consequences. Like the classic cartoon of the man looking for his keys only where the light shines from the lamp post, he explains how evaluation research on water and sanitation projects have focused almost exclusively on diarrheal morbidity while other fecally transmitted infections (less dramatic in their presentation but debilitating over the long term) are not so easily measured and may not feature at all, an absence that is inevitably repeated in nutrition and health care policy (International Initiative for Impact Evaluation, 2015 cited in Chambers, 2017). In effect, as the Big Push Forward claimed, knowledge as evidence becomes politicized: "For those who hold the purse strings, certain ways of knowing and assessing impact are considered more legitimate than others" (Institute of Development Studies, n.d.).

With experiences like these, methodological pluralism is increasingly being held up as an aspirational practice in development evaluation. Minimally, this is expressed in separate but equal terms: "While large quantitative studies are invaluable, rich qualitative descriptions of individual cases should have an equally prominent place in the evaluator's toolkit" (Picciotto, 2014, p. 9). More expansively, the integration of different methods is seen to reflect a deeper appreciation for complexity science and systems thinking as well as a way of exercising the moral obligation to employ a strategy more in keeping with a human well-being or justice agenda that embraces different ways of knowing. UNICEF's collection, *Evaluation for Equitable Results*, illustrates this trend (Segone, 2011). Rigor in this framing is as much about the relationships forged in the practice of evaluation as it is about the fit-for-purpose mixing of methods and paradigms (Chambers, 2017; Guijt, 2015; Patton, 2011; Picciotto, 2014). As quoted already, Jennifer argued for "engagement, interaction, listening and caring" (Greene, 2015, p. 218). For recipients of international donor assistance, this cannot come soon enough: "We have to put more heart than technique in this thing called development and external cooperation does not put heart into it" (Government officer, Ecuador, as quoted in Anderson et al., 2012, p. 30).

OPENING UP THE SPACE FOR METHODOLOGICAL PLURALISM IN INTERNATIONAL DEVELOPMENT EVALUATION: INTERSECTIONS WITH THE WORK OF JENNIFER GREENE

In this section, I explore how Jennifer's contribution can add to or enhance innovative practice in two areas touched on by this paper: methodological pluralism and the expectation of credible evidence in evaluation inquiry.

In a much-quoted study commissioned by the UK's Department for International Development (DFID), Stern et al. (2012) offer a broad range of designs and methods that "open up complex and difficult to evaluate programmes to the possibility of [impact evaluation]" (p. i). On mixed methods, they credit the influence of the early work of Greene, Caracelli, and Graham in outlining the various purposes for and benefits of mixing methods, but they also add a cautionary note:

> There remain questions for qualitative and interpretivist researchers about uncritical advocacy of mixed methods. For example: can data derived from different theoretical perspectives be combined? Might the data they produce be incommensurable, so the evaluator is left with adding up apples and pears? And what happens if different methods lead to different conclusions? (Stern et al., 2012, p. 30)

What would Jennifer say? Based on her contributions so far, I imagine she would counter this not with a fruit puree but a fruit salad in which these different flavors can be appreciated for what they contribute to the whole dish: a dialogue in which tensions in interpretations can be explored and decisions based on evaluation findings more thoroughly interrogated.

Given the immense challenge of proving results to demonstrate development or aid effectiveness, the field of development evaluation *is* moving forward, first by disrupting confidence in conventional procedures, then by creating space within the "results and evidence" agenda for methodological pluralism and a more democratized practice of evaluation. Many donor agencies and private foundations are now experimenting with mixing methods and participatory evaluation, even if it is not going far enough to bring paradigmatic positions into dialogue. This is the gauntlet Jennifer Greene has laid down.

Guijt (2015) asks, "Can a results and evidence agenda be used to the advantage of a transformational agenda? If so, how?" She argues that it can be if "feasible, useful, and rigorous... accompanied by autonomy and fairness, generat[ing] time and space for reflection on evidence of results, and...agile" (p. 194). Echoing Jennifer Greene, Guijt takes on the issue of rigor as having democratic and methodological dimensions. She reaffirms the importance of providing solid, actionable data but also argues for rigor "to be reclaimed beyond narrow method-bound definitions" and applied to "thought processes, methods, data, use, including voices that matter" (p. 195). It requires claiming power and agency over some of the processes earlier described as invisible control mechanisms.

There are several examples of development organizations that have claimed a degree of power and agency, challenging the audit culture of the aid industry and achieving some synergy between accountability and learning. For example, in the policy paper "Quality Before Proof," VENRO (2010), an association of 118 German Development NGOs, states the goals of impact monitoring to be learning, steering, accountability, and empowerment. Importantly, VENRO makes explicit demands of donors, politicians, and academics to understand and adjust expectations based on the principles and values of the NGO sector. Others have explicitly limited the extent to which its programs are based on linear theories of change with predetermined indicators, leaving space for an iterative process to unfold in sometimes unexpected ways, revealed through retrospective reflection and evaluation (Knox Clarke, 2017). Others position the decision-making agendas of different stakeholders (e.g., donors, intermediaries, community members) as having equal value, each demanding different methodological strategies (Mathie & Peters, 2013). Still others have negotiated with donors, or even justified their actions afterwards, and found the resources to show how number crunching to satisfy impact reporting requirements can be deepened by detailed learning from multiple stories of change, as with the Coady International Institute in Canada (den Heyer et al., 2021). What comes across clearly is a new dynamism in evaluation methodologies given multiple accountabilities, multiple channels for learning, and a language of partnership and solidarity (e.g., van den Berg et al., 2019).

Perhaps the strongest nudge comes from the urgency of global challenges as they stand today. In the face of climate change and a global pandemic—both contributing to ever-widening social inequalities—never has the demand for actionable evidence been stronger. The methodological, social, and political choices we make in evaluation will be judged by the pace with which such social inquiry informs and galvanizes appropriate action at every level. We need to reframe evaluation in the same breath as we reframe development, accountable to future generations.

CONCLUDING THOUGHTS

Reflecting on Jennifer Greene's contributions and their relevance to international development, I have tried to show that the disruptive effect of arguing for pluralism of method and paradigm, along with a commitment to democratizing evaluation practice, has been shaking things up for some time, with increased vigor as the challenges of global development become more urgent. This is why paying closer attention to Jennifer's challenge to bridge the gap between the theory and practice of mixing methods and paradigms is so important and timely. She has started to show how this can be done with real-world examples of evaluations that can be enhanced by different paradigmatic perspectives or different methods (e.g., Greene, 2015). I have also tried to show how her ideas about credible evidence provide specific guidelines for demonstrating rigor in applying values and methods. In that liminal space of uncertainty and confusion, stakeholders in evaluation are much more receptive to these ideas now than when she first expressed them.

This tribute to Jennifer has so far given minimal attention to her thoughts on the role of evaluators as the mediators among divergent evaluation stakeholders, and as proponents of methods appropriate for the evaluation task at hand. It is clear from her contributions that much is demanded of evaluators in terms of their skill sets and integrity, especially when unconventional or disruptive ideas come into play. Those of us who have been privileged to study and work with her know how high her own standards are in this regard, not only as an evaluator but also as an educator. Combining these, she has given us the chance to cross a threshold ourselves and champion evaluation as something of consequence.

NOTE

1. A larger swathe of development actors in recipient countries was included in a subsequent declaration (Busan, Korea, in 2011), when the agenda was broadened to refer to development effectiveness, rather than simply aid effectiveness.

REFERENCES

Anderson, M. B., Brown, D., & Jean, I. (2012). *Time to listen: Hearing people on the receiving end of international aid*. CDA collaborative learning projects. Retrieved from https://www.cdacollaborative.org/publication/time-to-listen-hearing-people-on-the-receiving-end-of-international-aid/

Chambers, R. (2017). *Can we know better? Reflections for development*. Practical Action Publishing.

den Heyer, M., Smith, E.W., & Irving, C. J. (2021). Tracing the link between transformative education and social action through stories of change. *Journal of Transformative Education, 19*(4), 421–432. https://journals.sagepub.com/doi/10.1177/15413446211045165

Eyben, R. (2015). Uncovering the politics of evidence and results. In R. Eyben, I. Gujit, C. Roche, & C. Schutt (Eds.), *The politics of evidence and results in International Development: Playing the game to change the rules?* (pp. 19–39). Practical Action Publishing.

Eyben, R., Guijt, I., Roche, C., & Schutt, C. (Eds.). (2015). *The politics of evidence and results in International Development: Playing the game to change the rules?* Practical Action Publishing.

Feinstein, O. (2019). Dynamic evaluation for transformational change. In R. D. van den Berg, C. Malgro, & S. Mulder (Eds.), *Evaluation for transformational change: Opportunities and challenges for the sustainable development goals* (pp. 17–39). International Development Evaluation Association.

Greene, J. C. (1994). Qualitative program evaluation: Practice and promise. In N. Denzin & Y. Lincoln (Eds.), *Handbook of qualitative evaluation* (pp. 530–544). SAGE Publications.

Greene, J. C. (2011). Values-engaged evaluation. In M. Segone (Ed.), *Evaluation for equitable development results*. UNICEF. Retrieved from http://www.clear-la.cide.edu/sites/default/files/Evaluation_for_equitable%20results_web.pdf

Greene, J. C. (2012). Engaging critical issues in social inquiry by mixing methods. *American Behavioral Scientist, 56*(6), 755–773. https://doi.org/10.1177/0002764211433794

Greene, J. C. (2013). Reflections and ruminations. In. D. Maartens & S. Hesse-Biber (Eds.), *Mixed methods and credibility of evidence in evaluation: New directions for evaluation, 138* (pp. 108–119). Jossey-Bass.

Greene, J. C. (2015). How evidence earns credibility. In S. I. Donaldson, C. A. Christie, & M. M. Mark (Eds.), *Credible and actionable evidence: The foundation for rigorous and influential evaluation* (Kindle ed.). SAGE Publications.

Greene, J. C., & Caracelli, V. J. (1997). Defining and describing the paradigm issue in mixed-method evaluation. *New Directions for Evaluation, 1997*(74), 5–17. https://doi.org/10.1002/ev.1068

Greene, J. C., Caracelli, V. J., & Graham, W. F. (1989). Toward a conceptual framework for mixed-method evaluation designs. *Educational Evaluation and Policy Analysis, 11*(3), 255–274. https://doi.org/10.3102/01623737011003255

Guijt, I. (2015). Playing the rules of the game and other strategies. In R. Eyben, I. Gujit, C. Roche, & C. Schutt (Eds.), *The politics of evidence and results in international development: Playing the game to change the rules?* (pp. 193–211). Practical Action Publishing.

Hall, J., Ahn, J., & Greene, J. C. (2011). Values engagement in evaluation: Ideas, illustrations, and implications. *American Journal of Evaluation, 33*(2), 195–207 https://doi.org/10.1177/1098214011422592

Institute of Development Studies (n.d.) The big push forward: Project. Retrieved from https://www.ids.ac.uk/projects/the-big-push-forward/

International Initiative for Impact Evaluation (3ie). (2015). *Water, sanitation and hygiene evidence gap map.* Retrieved from https://www.ircwash.org/resources/water-sanitation-and-hygiene-evidence-gap-map

Knox Clark, P. (2017). *Transforming change: How change really happens and what we can do about it.* ALNAP.

Mathie, A., & Peters, B. (2013). Joint (ad)ventures and (in)credible journeys evaluating innovation: Asset-based community development in Ethiopia. *Development in Practice, 24*(3), 405–419. https://doi.org/10.1080/09614524.2014 .899560

Organization for Economic Cooperation and Development. (2019). *The high level international fora on aid effectiveness: A history.* Retrieved from https://www .oecd.org/dac/effectiveness/thehighlevelforaonaideffectivenessahistory.htm

Patton, M. (2011). *Developmental evaluation: Exploring complexity concepts to enhance innovation and use.* Guilford Press.

Picciotto, R. (2014). Have development evaluators been fighting the last war . . . and if so, what is to be done? *IDS Bulletin, 45,* 6–16. https://doi.org/10.1111/1759 -5436.12109

Segone, M. (Ed.). (2011). *Evaluation for equitable development results.* UNICEF. Retrieved from http://www.clearla.cide.edu/sites/default/files/Evaluation_for _equitable%20results_web.pdf

Stern, I., Stame, N., Mayne, J., Forss, K., Davies, R., & Befani, B. (2012). *Broadening the range of designs and methods for impact evaluations.* Working paper #38. Department for International Development.

Van den Berg, R. D., Magro, C., & Mulder, S. S. (Eds.). (2019). *Evaluation for transformational change.* International Development Evaluation Association. Retrieved from https://ideas-global.org/transformational-evaluation/

VENRO. (2010). *Quality before proof* [Policy paper]. VENRO. Bonn, Germany. Retrieved from http://bigpushforward.net/resources

Whitty, B. (2015). Mapping the results landscape: Insights from a crowdsourcing survey. In R. Eyben, I. Gujit, C. Roche, & C. Schutt (Eds.), *The politics of evidence and results in International Development: Playing the game to change the rules?* Practical Action Publishing.

CHAPTER 12

IN PURSUIT
OF DEMOCRATIC VALUES

Transnational Influences
on Jennifer C. Greene

Melissa R. Goodnight
University of Illinois Urbana–Champaign

Cherie M. Avent
University of Illinois Urbana–Champaign

ABSTRACT

This chapter explores the meaning of Jennifer Greene's transnational and cross-cultural exchanges for her own development as an evaluator and for the evaluation field. Of keen interest is (a) how evaluation contexts and evaluators' work outside the United States shape Greene's understanding of evaluation and its larger social purpose and (b) how global, transnational, and cross-cultural exchanges matter to the present and future of the field. Our analysis searches for the meaning underlying Greene's participation in international exchanges about evaluation. First, through a close examination of her publications and then, by interviewing Greene, we explore how social

Disrupting Program Evaluation and Mixed Methods Research for a More Just Society, pages 165–177
Copyright © 2023 by Information Age Publishing
www.infoagepub.com
165

programs, communities, and evaluators beyond the Global North imaginary of mainstream evaluation have enriched her expansive vision of evaluation's relationship to making the world a better place.

I have been very lucky to have developed friendships and professional relationships with wonderful people in many parts of the world...I have been influenced by their values and their customs. I count my blessing that has happened. It's not that nobody does that kind of work in this country, but there's a cultural context in which my colleagues' evaluation takes place, and it's different than ours. I feel my thinking about evaluation, my sense of self as an evaluator, and my work have all been importantly influenced by a number of these relationships and friendships—because the good relationships were also friendships.

—Jennifer Greene, interview, 2021

The contributions of Jennifer C. Greene to the field of evaluation are innumerable. She has significantly influenced how evaluators think about the service of evaluation and how they put their values into practice within such service. While Greene's scholarship often conceptualizes these ideas in context of the United States, we wanted to understand her work and relationships beyond its borders. This chapter explores the meaning of Greene's transnational and cross-cultural exchanges for her own development as an evaluator and for the evaluation field, more broadly.[1] First, through literature review and then an interview with Greene, we sought to understand how the impact of values, evaluation contexts, and evaluators' work outside the United States progressively shaped her conceptualization of evaluation's possibilities and greater social purpose. We discovered that Greene's transnational exchanges have been a wellspring for her work's diversity and validation of ignored perspectives. Her friendships and evaluation experiences beyond the United States have nurtured her evaluator reflectivity, enriched her understanding of democratic values in evaluation, and altered her ways of doing evaluation. We have organized the chapter's sections, following the study design, to reflect the chronology of topics discussed and themes that emerged in our interview with Greene: (a) her foundational and enduring ideas about evaluation, (b) her engagement with values across countries and cultures, (c) her perception of U.S. connections to other evaluation contexts, and (d) her thoughts on the future of evaluation.

STUDY DESIGN

We employed two strategies, literature review and interview, in investigating Greene's transnational and cross-cultural exchanges to examine the following questions: (a) "How have evaluation contexts and evaluators' work outside the United States shaped Greene's understanding of evaluation and its larger social purpose?" and (b) "How do global, transnational, and cross-cultural exchanges matter to the present and future of the field?"[2] We reviewed Greene's published work on program evaluation to determine the publications that most represent her influential and lasting ideas regarding evaluation theory and practice. We opted to focus less on methodology and instead surveyed her substantive writing on (a) values "advocacy" or *values engagement*, (b) inclusive and diverse participation, and (c) democratic practices (e.g., deliberation). Next, we identified four publications in which Greene directly engages transnational or global themes, contexts, or audiences and used them to formulate our operating understanding of how she views cross-national and cross-cultural exchanges and their importance to evaluation's development and her own work (Greene, 2012, 2015, 2016; Mathie & Greene, 1997). Then, we had the privilege of interviewing Greene to discuss further some of her exchanges and opinions on the global, transnational, and cross-cultural character of the field and its future. We invited Greene to contemplatively discuss her transnational exchanges and collaborations with the twin aims of sharing these evaluation stories and reflecting on their personal value and broader implications for the field. As emerging scholars at the University of Illinois in the department where Greene crafted some of her most impactful scholarship on values-engagement in evaluation and mixed methods, we deeply feel her legacy's impact on the direction of our research and teaching, so it was a pleasure to further explore these topics with her.

FOUNDATIONAL AND ENDURING IDEAS ABOUT EVALUATION

Our conversation with Greene began with understanding ideas that have been foundational and unchanged throughout her career, recognizing they likely surfaced to some extent in her transnational experiences. She indicated that two topics have endured: evaluation as a service and the promotion of values. First, Greene noted that evaluation is "a service to the funder, the people running the program, and it should be a service to the people who are in the program." To her, while the service of evaluation can take many shapes, evaluation is easy compared to designing and implementing a program. Second, Greene discussed the role of values. Since her early writings, Greene has championed the idea that evaluators should be explicit about the value commitments undergirding their work (Greene, 1997, 2012, 2016; Greene et al. 2011) because, as she commented in our conversation, "evaluation makes judgments based on a selection of particular values." To not be explicit would be "disingenuous, even decep-tive to our audiences" (Greene, 1997, p. 28). In our review of Greene's writing, we noted consistency in her strategies for addressing values via "participation, voice, and engagement" with those who have a stake in the program (Greene et al., 2004, p. 99), yet we perceived a change in *language* describing her values: from pursuing democracy via multiple perspectives, inclusion, and pluralism (Greene, 1997; Mathie & Greene, 1997) to more directly challenging privilege, power, and authority (Greene et al., 2004). During our interview, Greene elaborated on this shift:

> In the earlier part of this [work], I certainly championed democratic values and thought about them in my work and tried to enact them. So, there was an intention there, but not necessarily a very good or thoughtful plan with specif-ics. Not necessarily [any] specific ways to make that happen like trying to hear from as many stakeholders as possible, being inclusive. There were broad ways to think about doing an evaluation that lots of people would call "democratic" in its process...At the beginning, they were these ideals, these abstractions that were part of my thinking and part of my work but kind of remained out there in the intellectual stratosphere. I think over time...I realized that this isn't just an academic framing of my work...So, when I really realized that in practice, then I think I became somewhat more intentional in thinking about "How do I advance fairness? What could I do with this interview activity, or this meeting, or this instrument to collect kids' perceptions of the program to make it as equitable, and fair, and inclusive as possible?" I'm not sure that would be the language I would have used, but it became more concrete after some period of time rather than a kind of vision...not just holding them as ideals but trying to bring them down to the actual day-to-day practice.

Apparent in Greene's comment is the shifting importance she ascribed to being deliberate in enacting values such as fairness or equity. She highlights that her evolution as an evaluator coincided with her revelation that values are not solely part of academic thought or discussion but require tangible actions in the conduct of evaluation, which must be governed by the evalu-ator's "genuine respect for the people in the context at hand."

ENGAGING WITH VALUES ACROSS COUNTRIES AND CULTURES

Next, we wanted to gain a sense of Greene's cross-cultural and global evalu-ation experiences. In "Culture and Evaluation," Greene (2015) proposes the benefit of transnational exchanges in deepening evaluators' practices within and outside the United States by adopting a "cross-cultural lens to offer insights into the practical challenges of conducting culturally respon-sive and responsible evaluation and to encourage a spirited international dialogue" (p. 93).[3] While the chapter offered critical reflections on her experience in Aotearoa, New Zealand, we wondered whether there were additional insights to be shared. We asked Greene to tell us ways in which colleagues or evaluation contexts outside of the United States have shaped her work. She stated:

> The most powerful example was getting to know a small group of Māori evaluators in New Zealand...I would go places where the Māori were leading the group and many of the kind of cultural customs of how one is with one another were very prominent...There was pride amongst the Māori...The Māori could do that the best, of course, because they're also Māori. They know the traditions, and they're part of the traditions.

In addition to recalling the *relational trust and acceptance* she observed between the Māori evaluators and communities (see Greene, 2015), Greene noted two impactful friendships with colleagues in Europe: Helen Simons from the United Kingdom and Tineke Abma from the Netherlands. Greene expressed about Simons: "I've been good friends with her forever. She worked...to develop and advance through practice a democratic way of thinking about evaluation...but it's distinctively British. Helen also used to do storytelling and poetry in her work." In speaking about Abma's influence, Greene explained: "She's incredibly articulate and visionary. She's always thinking about the next step, the next vision, or the next idea. She does so with great imagination and wonderful language. Tineke, she's just always there pushing against the next thing."

The Māori community, Helen Simons, and Tineke Abma have all importantly influenced Greene's thinking about evaluation and her role as an evaluator. Māori evaluators underscored the value added of communities or "insiders" shepherding evaluations to honor traditions and lived experiences. Helen Simons and Tineke Abma stimulated ways of thinking about and promoting democracy in evaluation that were innovative for Greene. Overall, there were clear commonalities amongst these cross-cultural relationships that shaped her views on engaging values and the role of evaluation in society. She explained:

> All these people I've mentioned, we also share a commitment to democratic values and to using our work to make it better for the people who are on the bottom. You can't have an evaluation of a program serving your neediest people when in that evaluation, the evaluators are looking down upon those people, when they're not advancing the values of generosity, grace, equity, fairness...that's what they do in their own distinct ways. It's not about elegance in evaluation, it's about using powerful and evocative evaluation as need be in service to the people in our society who are least well off.

The positioning of evaluators and evaluation in this quote is vital given that many evaluations assess programs aiding underserved, underrepresented, and historically marginalized populations. Her statement critically challenges evaluators' value commitments and enactments, particularly when the enacted values perpetuate pushing people to the margins. Further, she underscores the useability of evaluation to meaningfully attend to underserved individuals' or groups' needs. Taken together, her statements assert evaluation's potential to contribute to a better society.

Greene additionally described being inspired to try techniques she observed for generating, exploring, and sharing evaluation data. She pointed out literary strategies such as storytelling or role-playing as mechanisms that tended to promote stakeholder understanding of a situation and honored the values of the cultural context, connecting with "where people are."

> I was often inspired by how some of these evaluators thought about their work and the kinds of chances that they took. The Māori, Helen, and Tineke all made good use, and appropriate use, as best I know, of literary ways of doing and presenting their evaluation work. So, it'd be storytelling, or a short play. With their inspiration, I tried some of these things out over the years: We've had the superintendent of [a] school system...acting out a little skit, that were always data-based. They were crafted to really pinpoint a particular challenge or problem with teaching, or with the way classes are organized. The Māori people especially, but also Helen and Tineke, use literary ways of presenting findings to stakeholders because they could understand a lot better a story than they could [a] chart or a set of numbers. So, I also am very appreciative of that.

Another example of the influence of Greene's transnational interactions on her personal theorizing and practice is Greene's deferential evaluator stance within unfamiliar cultural contexts:

> I think I was situating myself as a learner of value systems in other countries...I think evaluation is the scientific, systematic practice of valuing, so we have a legitimate place to question values but it's also extremely delicate. I don't think I would have ever felt empowered to impose a set of values on a context that did not advocate for them. That's not my job. That would have been way too invasive. In some cases, we would negotiate the values that would be present: *It's not our job to impose values that are not welcome in that context.*

Greene's remarks illustrate two considerations. First, what role(s) should an evaluator assume when working in cross-cultural and transnational spaces? This consideration is important given the complex, sociopolitical spaces in which evaluations occur and the history of marginalization experienced by communities. Second, there's awareness of evaluation as a practice that encompasses value commitments that must also be balanced with the values and needs of the context. As Greene iterates, while there may be negotiation, it's imperative that unwanted values not be forced upon people who have an interest in the program.

CONNECTING THE UNITED STATES WITH OTHER EVALUATION PRACTICE CONTEXTS

Mathie and Greene (1997) examined two evaluation contexts—one in the United States and one in Papua New Guinea—analyzing the tension between increasing diversity in engagement and facilitating transformative evaluation action. In the 2 decades that have passed, we were curious about how Greene saw evaluation outside the United States as impacting thinking and practice. She remarked, "I don't think it has influenced it very much." Her reasoning was evaluation's institutionalization across countries is diverse. The exchanges that happen within the formalized field of evaluation (e.g., its journals and conferences) are not necessarily institutionalized in the evaluation policies and work of organizations that fund and carry out evaluations. However, her response was, "Oh, absolutely, absolutely . . . I think there's a lot we can learn" when asked if there were things evaluators were doing outside the United States that could be productively incorporated into domestic evaluations. She highlighted again how both Māori and non-Māori evaluators (Pākehā) had accomplished meaningful evaluation work within Māori communities by learning and adopting "the Māori way." The depth of these evaluators' cultural competence has supported the growth of "culturally responsive and responsible" programs supporting the broader cultural revitalization of Māori across the country (Greene, 2015, pp. 97–99). Greene explains that such apprenticeship in learning how to interpret and represent an underserved community's experiences and elevate their interests in evaluation is terribly important. Reflecting on her own evaluation projects in the United States, Greene expresses that such understanding is something that she has occasionally found impenetrable in her efforts to cultivate a deep sense of people's experiences:

> I did some work for a few years at [a] correctional facility, I volunteered as an evaluator . . . I never understood what it's like to be a prisoner. I tried. It was all men, and I was [an older woman] . . . I just thought I can't understand this by my occasional interview, my occasional visit, and even [asking] "tell me about your day yesterday," or "pick a day that was kind of interesting." We used a lot of qualitative methodologies. I wanted to know what it was like to be there because I think that gives you some insight into the meaning of the words that you're hearing, but my experience basis was just not broad enough to capture that. We did a lot to feature the voices of the men in the Center, who were part of the program, rather than us. You know, there are limits to what you can do

as an evaluator. I fear that sometimes we think we know; we think we understand the context; we think we know what it's about, what the norms are, and we might be really wrong. And it was clear, I didn't know. But other times, I might assume that I know. And I don't, or I'm mistaken in my knowledge.

Knowing what it is like to occupy marginalized positions within societies can be extremely delicate and difficult for an outsider, but such insights are essential to evaluate programs effectively. Greene's statement emphasizes four issues—a commitment to knowing, humility, methodological strategies, and time—in gaining a deeper cultural competence and knowledge of the life-worlds of marginalized individuals when an evaluator occupies a social standpoint beyond those realities. In sum, what Greene identified as a strength of transnational evaluation engagements for U.S. evaluators are examples of cultivating a deep empathy for communities that societies tend to demean, misunderstand, and misrepresent in evaluation and ultimately underserve in social programming.

THE PRESENT AND FUTURE OF EVALUATION GLOBALLY

In Greene's 2016 chapter "Advancing Equity," she highlights "the now-global evaluation community" in her assertion that it is "past time . . . to begin naming and claiming the values embedded in our evaluation practice" (p. 56). As a follow-up to that idea, we inquired how the global or transnational orientation of evaluation has altered her recent thinking about values engagement and how she perceived the universality versus cultural specificity of values that guide evaluation today. Greene made three points in her response: (a) there is a difference in contemplating values in evaluation within democratic versus nondemocratic countries, (b) evaluators in their own national contexts should feel fully empowered and responsible for promoting democratic ideals, and (c) evaluators can play a potentially supportive (not authoritative) role in furthering democratic ideals when they are outsiders. Elaborating on the issue, Greene remarks:

> I think mostly if I've done any work abroad, it's been in democratic countries. I think in democratic countries, certainly in our own country, we should not only feel absolutely free, but we should take it upon ourselves to advance democratic values as the criteria upon which we should judge the goodness of a program. I mean there may be others [values], but fundamentally, a program, especially that's publicly funded, should advance equity and fairness and opportunities for all.

Moreover, in the United States, Greene reemphasizes her earlier point that the value of equity is neither an optional commitment for evaluators nor a principle that remains abstracted from practice—rather, it is a fundamental responsibility that informs evaluator actions and evaluation processes.

> I feel totally comfortable—in fact, it's our *duty* to advance equity as a part of the criteria upon which we make judgments, and it's not just criteria, you have to gather data from the full range of people who are doing the program, who are participating in the program. You can't sample just one group, so equity has to permeate *all* of what we do as best we can, knowing there's always limits on that, but limits aren't an excuse for not doing it.

So, value engagement with equity shapes criteria for judging the program, the selection and participation of participants, and the decisions regarding data collection (Greene, 2012). Engagement with values is interwoven with "evaluative thinking and judging" (Greene, 2012, p. 198). However, as far as advocating equity or other democratic values in contexts outside of the United States, Greene's sense of evaluation authority shifts while her own evaluation values do not:

> I would not do that [assert values] in another country myself; I would support my colleague from that country to do that, but I think that's out of balance. Even [in] the other democratic countries, I just don't know the norms well, and so I might simply [say], "I would like to use these criteria to make judgments of the quality of the program, are there others that you would like to add?" But in this country, no excuse for not using them.

Greene emphasizes her insufficient cultural knowledge and her outsider position in her discomfort with asserting values beyond the United States. Yet she shares strategies for how she might support democratic values engagement in another context: respectfully suggesting and reflectively questioning criteria for guiding a program evaluation and encouraging insider colleagues to pursue the equity-focused or democratic values that they identify as appropriate for their circumstances. Greene expands her explanation later in our conversation.

> I think I would turn to the natives of a country to see if criteria like equity, fairness, and respect are appropriate criteria for judging the quality of the work that is done in [their] program. So, I would work with ... my colleagues from that country to try to persuade them that these are good ideas, that these are good criteria. I would add something like equity to an existing set of values unless there was real danger in doing that. But, I think if the RFP for evaluation proposals or something articulated a set of values, I would follow those and—maybe not using provocative words like "equity," but perhaps, it would depend on the context—I would feel committed to doing what the funder of the program and the evaluation wanted. I'm not an international rebel rouser. I know a little bit about some other countries, but I think it would be

disrespectful, it would be arrogant and disrespectful to claim, "Well, we do it this way in the United States, so you should do it this way here." No.

Greene's comments reflect pragmatism and the delicacy of being responsive to and respectful of a particular cultural context while also engaging values that can further equity and democracy—as conceptualized and afforded in that environment.

In closing our interview, we asked Greene her thoughts about the globalization and direction of evaluation over the next 5 to 10 years. In her response, Greene spotlights two issues: (a) the role of American evaluators in the field's continued global development and (b) the need for fostering a reasonable view of evaluation's contribution to societies. Regarding American evaluators' leadership in the field and building coalitions to advance democratic values in evaluation, Greene recommended caution and adopting a pathway of reflection, grace, and commitment to openly deliberating values.

> I think we need to get our own act together in the U.S....identifying the many ways that evaluation is done in this country and whose interests get served by each of those...I think there's a significant group of evaluators who agree...that this evaluation should not simply ask what the funders want to know. That's not democratic...I could see a small group of international evaluators committing to try to do this [too]...but I can't see any global movement because I think evaluation [is] institutionally positioned in different places in different countries, so I think there could be a group here in the U.S. that does this more vocally, and more aggressively, and with more publications...But I think we just do it step by step.

Greene identifies "a significant group of evaluators" in the United States who emphasize values in evaluation and are responsive to diverse communities, yet she does not view evaluation practice in the country as a model for other nations. This raises questions about the logic and justice of the United States—and other Western contexts—serving as default political and cultural environments for the field's standard setting and professional statements. Greene asserts that evaluation's growth has generally benefited individuals and societies worldwide. Nevertheless, she acknowledges evaluation's limited capacity to be the main mechanism of equitable social change globally.

> Globalization [of evaluation] is a good thing in many ways...I think it's a good thing for governments and communities and people doing the hard work on the ground...But evaluation is not going to save the world...I think we want to be appropriately modest in our aspirations. I think of the people who save the world as the frontline protesters and those who are organizing labor unions. I mean, the people who are working with and for the people, so I don't see ours as a radical [act]—it could be a radical activity in some contexts but that would be for people who live in that context to decide. I think

evaluation champions values like equity and fairness. And using data and being reflective. There's a whole host of great things that we do as human beings
that I think are represented in our craft. I certainly think evaluation can contribute to activities and programs and endeavors that are struggling to make
the world better, but I don't think we take the lead. I think we're the sidekick.

In elevating the direct struggles of citizens striving to "make the world
better," Greene distills a respectful evaluator disposition that responsibly
serves democratic values. This respectful evaluator is "appropriately modest" about one's contributions; champions "values like equity and fairness" while "using data and being reflective"; and acknowledges the labor
and sacrifices of real changemakers who work "with and for the people"
through programs, unions, political organizing, and protest. In her humble
and contemplative, yet bold, scholarship, Greene's thinking and practices
are an inspiring representation of "the great things we do as human beings" in our craft of evaluation. We see Greene as one embodiment of the
ideal evaluator-learner that she describes. She encourages us to "get out
there in the world" and find more evaluators who inspire us with their practices and challenge our thinking.

CONCLUSION

Our review of Greene's literature and interview aimed to understand the
meaning of her transnational and cross-cultural exchanges for herself and
the field broadly. What we have found is, first, Greene's friendships with
evaluators outside the United States contributed to her reflectivity as an
evaluation practitioner and theorist. The contexts of these relationships
and experiences placed her positionality front and center, thus making
it essential to adopt a learner perspective. Additionally, the relationships
enhanced her evaluation strategies for connecting people to stories or information that resonated with them. Second, Greene's explication of the
criticality of values engagement to evaluators' work does not mean she believes that evaluators claim value authority. Rather, following the educative
mission of evaluation (Greene, 2013, 2016; Greene et al., 2004), evaluators
should seek to be astute learners of values in the contexts and communities
that they cannot claim as culturally and nationally their own. Our interview
with Greene also suggests that even values such as equity, justice, respect,
and fairness are contextually bound. Therefore, the enactment of each may
vary depending upon community customs and needs.

Her remarks also raise considerations for the field as it continues to
evolve both inside and outside the United States. First, there's a continued
need for evaluators to explicitly communicate their value commitments
to stakeholders. How evaluators design and facilitate evaluation processes

reflect their value commitments whether they acknowledge them or not. Second, increased opportunities for cross-cultural and transnational dialogue about the complexities of values and what it means to engage them can provide space for evaluators to critically reflect on their theoretical and methodological choices in practice. Third, given the globalization of evaluation, there is a need to continue working for its pluralistic conceptualization and transnational institutionalization, particularly in how standard setting, professional association leadership, and disciplinary journals and publishing can become more inclusive of the diversity of contexts and cultures that are truly affected by the field and should be represented within it.

NOTES

1. Greene's exchanges are evident across various internationalized publication and professional meeting venues (e.g., *American Journal of Evaluation*, Information Age Publishing's Evaluation and Society book series, European Evaluation Society, South African Monitoring and Evaluation Association, and Center for Culturally Responsive Evaluation and Assessment conferences) and deepened by sustained friendships.

2. We use the term *global* to refer to phenomena that are not specific to identifiable countries but are operating at a meta level beyond country borders whereas *transnational* refers to phenomena that cross the borders of two or more countries and are identifiable to those countries. We can understand transnational phenomena as having defined versus ubiquitous contexts.

3. While this chapter focuses principally on the impact of transnational exchanges and friendships in Greene's evaluation work, this quote highlights the substantial influence of evaluators who developed culturally responsive evaluation (CRE) in the United States (many of whom have been Greene's university colleagues) like Stafford Hood, Melvin Hall, Denice Hood, Karen Kirkhart, and Rodney Hopson. Specifically, the language "responsive and responsible" is borrowed from CRE and explained by Hood et al. (2015): "Being 'responsible' is viewed as an active behavior manifested in advocacy of social justice for those who had been traditionally disenfranchised. To act on this responsibility requires one to be responsive by being aware and recognizing the centrality of culture and cultural context in our evaluative work and identifying the appropriate methods and tools that will best serve the community" (p. 309, drawing on Hood & Hall, 2004).

REFERENCES

Greene, J. C. (1997). Evaluation as advocacy. *American Journal of Evaluation, 18*(1), 25–35. https://doi.org/10.1016/S0886-1633(97)90005-2

Greene, J. C. (2012). Values-engaged evaluation. In M. Sergone (Ed.), *Evaluation for equitable development results* (pp. 192–2017). UNICEF.

Greene, J. C. (2013). Making the world a better place through evaluation. In M. Alkin (Ed.), *Evaluation roots: A Wider perspective of theorists' views and influences* (pp. 208–217). SAGE Publishing.

Greene, J. C. (2015). Culture and evaluation: From a transcultural belvedere. In S. Hood, R. Hopson, & H. Frierson (Eds.), *Continuing the journey to reposition culture and cultural context in evaluation theory and practice* (pp. 91–108). Information Age Publishing.

Greene, J. C. (2016). Advancing equity: Cultivating an evaluation habit. In S. Donaldson & R. Picciotto (Eds.), *Evaluation for an equitable society* (pp. 49–66). Information Age Publishing.

Greene, J. C., Ahn, J., Boyce, A., Hall, J., Johnson, J., & Samuels, M. (2011). *A values-engaged, educative approach to evaluating STEM education projects*. University of Illinois at Urbana-Champaign.

Greene, J. C., Millett, R. A., & Hopson, R. K. (2004). Evaluation as a democratizing practice. In M. T. Braverman, N. Constantine, & J. K. Slater (Eds.), *Foundations and evaluation: Contexts and practices for effective philanthropy* (pp. 96–118). Jossey-Bass.

Hood, S., & Hall, M. (2004). *Relevance of Culture in Evaluation Institute: Implementing and empirically investigating culturally responsive evaluation in underperforming schools*. National Science Foundation, Division of Research, Evaluation, and Communications, Directorate for Education and Human Resources (Award #0438482).

Hood, S., Hopson, R. K., & Kirkhart, K. E. (2015). Culturally responsive evaluation: Theory, practice, and future implications. In K. E. Newcomer, H. P. Hatry, & J. S. Wholey (Eds.), *Handbook of practical program evaluation* (4th ed.; pp. 281–317). Wiley.

Mathie, A., & Greene, J. C. (1997). Stakeholder participation in evaluation: How important is diversity? *Evaluation and Program Planning, 20*(3), 279–285. https://doi.org/10.1016/S0149-7189(97)00006-2

SECTION III

MEANINGFUL DIALOGUE: DISPOSITIONS, IMPOSITIONS, AND ASSUMPTIONS MATTER

Section III showcases how Greene has disrupted social inquiry through the dialogic aspects of her work. Building on Greene's work on dialogue in democratic evaluations, in Chapter 13, Wenjin Guo, Leanne Kallemeyn, Elissa W. Frazier, and Eleanor N. Titiml explicate the dispositions and conversational processes evaluators use in dialogue to support evaluation and a just society. Chapter 14 by Jill Chouinard reframes dialogue, focusing on linguistic dimensions of dialogue (i.e., power and voice) and how they matter for interpersonal communication and broader interconnected culturally diverse spaces and contexts. In Chapter 15, through a dialogic exchange, Sarita Davis and Alexis Kaminsky share how Greene supported their thinking as they questioned issues related to the imposition of Eurocentric ideas, race, gender, power, and privilege in evaluation practice.

CHAPTER 13

MICRO-LEVEL PROCESSES OF DIALOGUE FOR TRANSFORMATIVE EVALUATION PRACTICE AND A JUST SOCIETY

Wenjin Guo
Loyola University Chicago

Leanne Kallemeyn
Loyola University Chicago

Elissa W. Frazier
Loyola University Chicago

Eleanor Ngerchelei Titiml
Loyola University Chicago

Disrupting Program Evaluation and Mixed Methods Research for a More Just Society, pages 181–197

ABSTRACT

Inspired by Jennifer C. Greene's scholarship in democratic evaluation, re-lationships, and dialogue within evaluation practice, we discuss micro-level processes in dialogue within evaluation practice for a just society. We begin by outlining Jennifer's contributions of advocating for democratic values and promoting inclusive dialogue. We describe our community of practice and share how we engaged in a safe, inclusive space that allowed us to discuss our experiences, inquiries, and ideas regarding dialogue in evaluation practice. Our thoughts and conclusions of micro-level processes of dialogue are high-lighted. We recognize dialogue as a form of reflection, self-care, and building trusting communities. Additionally, dialogue is also a tool for understanding differences and communicating shared values among stakeholders, especially when power differences and diversity of culture, ideas, and beliefs exist in the evaluation space. Finally, dialogue is a form of resistance or self-preservation that is manifested when speaking up to address tension or discomfort. It is also manifested when choosing to remain silent in order to balance power dif-ferences or refrain from exacerbating problematic norms. Overall, we empha-size that dialogue for a just society, reflective of the various micro-processes we describe, is a seed for change. It exists in different forms, with each form pos-sessing catalytic potential. When sown with intention and fostered with care, dialogue blooms to inspire something unique and unexpected. This mirrors our experience in our community of practice, where authentic dialogue was central and supported by trust and care. As a result, we grew and changed, as did our evaluation practice. Dialogue, as a seed for change, is transformative in nature. Our experience reaffirmed our conviction that dialogue in evalua-tion practice, although emotionally laborious, is justice work and, therefore, important and necessary.

* * *

I plant my feet into the earth, breathe deeply, and slowly scan the crowd around me.

I close my eyes to feel their presence and listen to their hopes, desires, and what is most important to them.
On one hand, we are planning our next steps. On the other hand, we are waiting for permission from each other to walk next to each other, to see each other, not as the positions we hold, but the way we care for each other.

They must gesture for me to come a bit closer. They must invite me into their conversations to talk and walk with them.

Although my voice is welcoming, there are always a few gasps when I speak.

Whether they think they are ready or not, no time is ideal for the questions I ask or the ground we break together. I ask them to pause, to move with purpose. Our relationship requires them to consider their ways, our way

together, and the way ahead. I am always invited to their gatherings, but I
am not always asked to speak.

    —Elissa W. Frazier, inspired by a Chinese idiom, 鹤立鸡群,
    pinyin: hè lì jī qún, like a crane standing among roosters, and
    our experiences with dialogue in evaluation practice

The purpose of this chapter is to offer how we, inspired by Jennifer's scholarship, purposefully use dialogue for a just, democratic society in micro-
level processes of evaluation practice. Jennifer recognized the significance
of bringing together different perspectives in evaluation practice. We revisit Jennifer's contributions, including centering relationships in evaluation practice, the significance of dialogue in democratic evaluation, and
her advocacy for historically marginalized communities. Next, we describe
our community of practice, rooted in a transformative framework, followed
by our insights and narratives illustrating three micro-level processes of dialogue in our evaluation practice: reflection, care, and trust; understanding
difference and communicating shared values; and resistance. We have intentionally embedded alternative forms of representation throughout this
chapter, including poems, metaphors, stories, and a graphic. They were
central to understanding our differences in our community of practice. We
conclude this chapter with a metaphor of how dialogue is a seed of change,
representing our transformative learning and authentic hope for a just society through dialogues.

JENNIFER C. GREENE'S CONTRIBUTIONS:
ADVOCATING FOR DEMOCRATIC VALUES AND
PROMOTING INCLUSIVE DIALOGUE

Jennifer's scholarship provided a foundational vision for evaluation practice. Amid evaluation norms for value-neutrality and methods-centrism,
her evaluation practice was rooted in democratic evaluation, dialogue, and
advocacy. Jennifer legitimized practicing evaluation from an explicit value
commitment for democratic pluralism (Greene, 1997). She did so syner-
gistically with colleagues offering similar, alternative approaches to evaluation, including but not limited to Guba and Lincoln (1989), Fetterman et
al. (1996), Mathison (1996, 2000), Abma (1997, 1998), Schwandt (1997),
House and Howe (1998, 1999), Mertens (1999), Rallis and Rossman (2000),
and Hood (2000, 2001).

    Although not without being contested (Stake, 2000), Jennifer (Greene,
1997) advocated for promoting democratic pluralism, meaning evaluation
serves as a "force for democratizing public conversations about important
public issues . . . with democratic principles of equality, fairness and justice

as guides" (p. 28). At its heart, democratic evaluation involves broad inclusion of stakeholders in the evaluation process and facilitating dialogue about values among stakeholders. Thus, Jennifer intentionally invited stakeholders with different perspectives into the same spaces for respectful dialogue and learning.

Jennifer emphasized the centrality of relationships in evaluation practice (Greene, 2001, 2005, 2015) as foundational for the practice of democratic evaluation. For Jennifer, relationships were "about power and voice, agency and moral purpose, caring and empathy, understanding and acceptance of difference" (Greene, 2001, p. 182). She recognized that programs we evaluate are built on relationships; thus, the nature of relationships within an evaluand deeply matters to program quality and effectiveness. Finally, she emphasized that meaningful and effective dialogue depended on "relationships of trust, respect, caring, openness to difference, and political relationships of commitment to equity of power and voice" (Greene, 2001, p. 182). Jennifer was present on-site in her evaluation projects, modeling this relationship building for other stakeholders.

Central to democratic evaluation is dialogue. Jennifer (Greene, 2001) referred to evaluative dialogue as "engaged, inclusive, and respectful interactions among evaluation stakeholders about their respective stances, values, perspectives and experiences, dreams and hopes, and interpretations of gathered data related to the evaluand and its context" (p. 182). Such dialogue necessitates intentionally inviting stakeholders with different perspectives into the same space to engage with each other. The purpose of such dialogue is to foster "understanding and respect, though not necessarily agree with, one another's perspectives" (p. 182), which in turn "engender more reciprocal, equitable, and caring stakeholder relationships" (p. 182). Thus, "genuine dialogue is meaningful evaluation" (p. 184). Consistent with Jennifer's (Greene, 1997, 2000) scholarship, we aim to root understandings of dialogue intimately with our evaluation practice in this chapter. In her work, Jennifer found that alternative forms of representation, such as poetry, performance, and visual displays, facilitated the engagement of alternative perspectives (e.g., Johnson et al., 2013), as we do.

As a scholar-practitioner, Jennifer (Greene, 2000) openly acknowledged the challenges of advocating for democratic values, fostering dialogue in evaluation practice, and building relationships amid power imbalances and historic inequities. She acknowledged power and privilege and the transformative potential of evaluation. She also wrestled with fostering inclusion and dialogue in her practice. For example, Jennifer (Greene, 2000) encountered several challenges in a 3-year democratic evaluation of a new high-school science program. Significant stakeholder voices from parents and community program supporters were usually left out of the evaluation. Although democratic evaluation aims to create space for dialogue on multiple values, dominant voices centered dialogue on methods rather than on values. In the final year, the evaluation team experienced limited authority to influence policy when the local school board made decisions on the science program without having the final report on the evaluation findings. Such struggles are not unique to Jennifer's work (see also Abma, 1997; Hopson et al., 2012; Ryan et al., 1998), nor are they unique to the struggles of addressing systemic inequities.

In addition to Jennifer's substantive contributions, her scholarship for the intersection of evaluation and advocacy has opened a door for a new generation of evaluation scholars. These scholars have embraced advocacy and argued for explicitly addressing oppression and privilege as central to evaluation practice (e.g., Boyce, 2019; Chouinard, see the following chapter; Guajardo et al., 2020; Hall, 2020; Lemos & Garcia, 2020; McBride et al., 2020; Reid et al., 2020). They have also embedded their work in advocating for value, such as social justice and addressing inequities. More significantly, the work of these scholars also demonstrates the importance of evaluators engaging in critical self-reflections (e.g., Hall, 2020; McBride et al., 2020; Reid et al., 2020).

OUR COMMUNITY OF PRACTICE

I (Leanne) met Jennifer after I had recently relocated to Champaign-Urbana, Illinois. I inquired about the PhD in educational psychology. Jennifer invited me into her home (not her office). We talked about my career aspirations, experiences with research and evaluation, and concerns I had with applied research after working on a randomized trial of a home visiting program. She probed to learn about the qualitative follow-up study I assisted with to explore the nonsignificant findings. Her interests intrigued me.

As an undergraduate psychology major who was cultured in assumptions of post-positivism and whose life experiences did not mesh with these assumptions, I appreciated Jennifer's engagement with values, politics, power, and privilege in evaluation courses and practice. I worked with Jennifer on a mixed-methods analysis in an evaluation project (Greene, 2007, p. 160), in which I was one of three graduate students responsible for analyzing a data source in the evaluation. We each did so independently and then developed a joint display for combining our analyses. I experienced the value of juxtaposing these different perspectives when the entire evaluation team had generative, rich dialogues regarding discrepancies, which led to new interpretations and understandings.

In her own scholarship, Jennifer worked with other women scholars/evaluators and scholars of color, supporting their work and contributions to the field of evaluation. In the process, she affirmed the broadening of voices and perspectives in the field of evaluation and mentored the next generation of evaluators. So as a tribute to Jennifer, I wanted to do the same. I invited doctoral students with different perspectives and strong interests in evaluation to share in this work. We are from diverse cultural, linguistic, educational, and professional backgrounds, which brings alternative lived experiences to extend the breadth and depth of our meaningful dialogues. In Fall 2020, we formed a community of practice (Wenger, 1998, 2010) that served as a space to inquire regarding how to engage in dialogue during evaluation practice. Our practice is grounded in transformative learning (Mezirow, 1991, 1997) to explore our multiple identities and negotiate meaning making regarding the micro-level processes of dialogue in evaluation practice.

We have different identities, subjectivities, and backgrounds; together, we collaboratively created a safe, welcoming, and respectful community. Leanne identifies herself as a White, associate professor, mother of two children, cisgender female. She teaches evaluation courses, mentors students, and has served on multiple evaluation projects of educational programs. Elissa identifies as a Black woman in America, a mother of four, an educator, an evaluator, and an emerging culturally responsive researcher. Wenjin and Eleanor are both female international doctoral students from the People's Republic of China and the Republic of Palau, respectively. Wenjin engages in university–community partnerships to support students in under-resourced communities. Eleanor has evaluation experiences in higher education, college-access programs, government, nonprofits, and community organizations in the United States and her home in Palau.

Our Process: Reflective Learning in Practice

Central to defining a community of practice are shared norms. Our norms included (a) to practice authenticity with awareness of self and self in relation to others; (b) to conduct value inquiry; (c) to engage in unlearning values, assumptions, beliefs, biases, and so on that we bring; (d) to attend to positionality, including the boundaries we span as insiders and outsiders of communities; and (e) to aim for transformation and change. We acknowledged our differences and the power balance, candidly engaging in those dialogues since our first meeting. As our relationships developed, we found ourselves feeling more comfortable and confident in having different perspectives, unlearning through priorly held assumptions and misconceptions because of the caring and respectful community we had created. Though we might have disagreements, we respect, understand, and acknowledge the differences.

We engaged in pre- and post-meeting individual reflection journals and joint reflexive dialogues, using Pine's (2009) models of reflection to guide journals and reflexive dialogues. Within our community of practice, we focused on wider social and ethical issues, such as our roles, values, multiple identities, positionalities, assumptions, and follow-up actions to promote dialogue in evaluation for a just society. We aimed to unlearn and reconstruct our understandings of different evaluation projects, evaluation practices, and multiple identities.

Our vigorous and competent engagement in those dialogues was the groundwork of Wenger's (2000, 2010) community of practice. Based on "collectively developed understanding of what [our] community is about," we "hold each other accountable to this sense of *joint enterprise*," welcome mutual engagement, and develop "a *shared repertoire* of communal resources" (Wenger, 2000, p. 229, emphasis in original). Also, we practiced learning as the production of social structure (Wenger, 2010). First, we participated in "conversations [and] reflections" (p. 180). Second, we crafted "physical and conceptual" artifacts, as poems, visualizations, stories, and a book chapter that "reflect our shared experience" (Wenger, 2010, p. 180). More importantly, the community of practice we participated in echoed Jennifer's advocacy to include female and marginalized scholars and use alternative forms of representation.

Our Disruptions for a More Just Society

We purposefully went beyond forming a holistic understanding of evaluation practice to intentions for a more just society. Mezirow (1991) described the process of transformative learning as:

> An enhanced level of awareness of the context of one's beliefs and feelings, a critique of their assumptions and particularly premises, an assessment of alternative perspective, a decision to...take action based upon the new perspective, and a desire to fit the new perspective into the broader context of one's life. (p. 161)

We supported each other to "become aware and critical of [our] own and others' assumptions...practice in recognizing frames of reference and using [our] imaginations to redefine problems from a different perspective" and "participate effectively in discourse" (Mezirow, 1997, p. 10).

Further, we brought transformative learning to the workplace and the community to practice and "seek a transformation at the community or societal level" (Merriam & Bierema, 2014, p. 94). We also assisted one another to become "more autonomous thinker[s]" by deconstructing and reconstructing our values, meanings, and purposes through "critical reflection, awareness of frames of reference," and engagement in different contexts (Mezirow, 1997, p. 11). In sum, we transformed through critical reflection practices, caring and authentic relationships, and a habit of mind to disrupt systemic biases and stereotypes.

MICRO-LEVEL PROCESSES OF DIALOGUE

Jennifer (Greene, 2006) referred to the micro-level of evaluation practice as the relationships that evaluators build and the processes and interactions that enact these relationships. We provide another lens centering on internal processes and dialogue within oneself while engaging with external stakeholders or evaluation team members. In our experience, we come into evaluation spaces with all of our humanness (McBride, 2015); our familial, cultural, and community experiences; and our values and beliefs—we come in as our whole selves. However, there are times when we choose which identities to bring to the fore with an awareness of how those identities are weighed positively or negatively in a social context, thus opening doors or creating barriers.

Our evaluation experiences intertwined with our identities provide us with membership, helping to build trust and empathize with some while distancing us from others where we share fewer associations. As members of nondominant communities (e.g., Palauan, Black, Chinese), we (Eleanor, Elissa, and Wenjin) posit that in seeking to build external relationships and engage in evaluation practice, there are decisions that are made internally that help us to position ourselves to cultivate dialogue where it is needed. Our internal dialogue was necessary for us as more than reflection as it yielded in-the-moment outcomes as decisions or directional pivots that we might not have made otherwise. Furthermore, our discussion of internal dialogue centers on the internal tension we felt at times to make the best choice for the situation as racially, ethnically, or linguistically minoritized individuals who happen to be serving in the role of evaluators. Additionally, we are all women, adding a gender intersection of what it means to exist in the field of evaluation in that capacity.

Figure 13.1 is a visual representation of the inner dialogue and microprocesses individually (not collectively) engaged within our evaluation practice across our different contexts. This figure demonstrates the microlevel processes that were central in our discussions and perhaps not emphasized in prior scholarship. We chose to focus on these in an effort to deepen Jennifer's scholarship. Our conversations yielded three themes that encompassed purposes for dialogue in evaluation. Our dialogues allowed us to contend with how we could or should maintain the status quo or contribute to a shift in power dynamics toward a more inclusive, empathic, or agentic

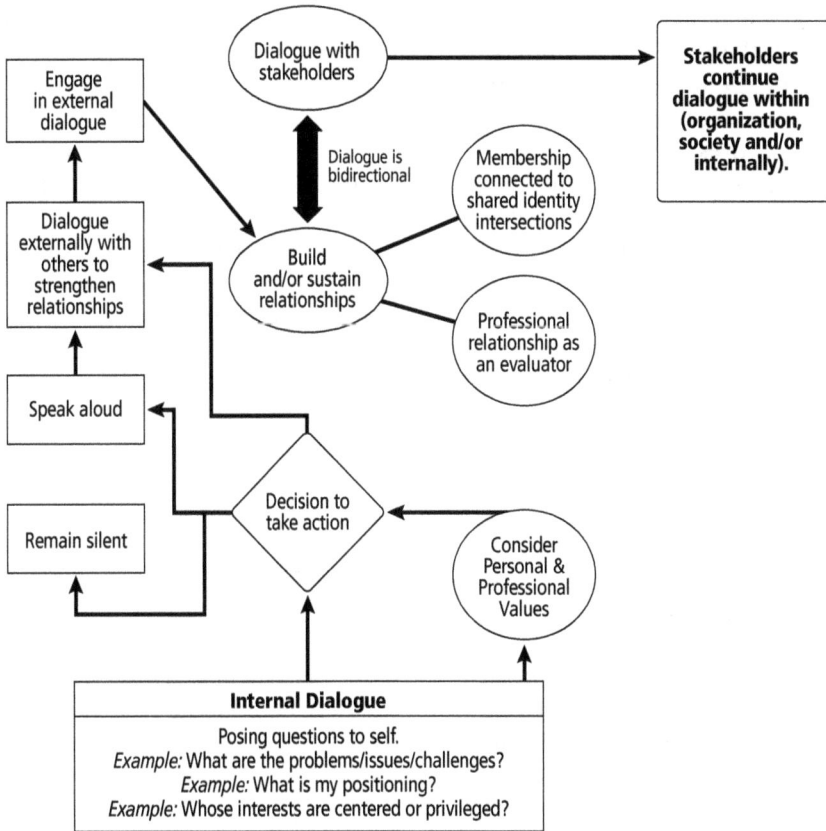

Figure 13.1 Micro-level processes of dialogue in evaluation practice.

space. Out of these dialogues came an action, a decision to remain silent, or a resolve to speak aloud. Often, though not always, the process of dialogue led to interpersonal dialogue and subsequently strengthened relationships.

Further, we discussed the role of internal dialogue as central to our ability to navigate challenges within evaluation teams and with stakeholders. We used internal dialogue to consider when and how to confront situations where power imbalances emerged and caused tension. These intrapersonal dialogues were intimately tied to our roles and our social and cultural identities. There is a Chinese idiom, think thrice before acting (in Chinese: 三思而后行, *pinyin*: sān sī ér hòu xíng). Three times can be understood as indefinite, meaning to think and reflect multiple times. Having internal dialogues is to engage ourselves to think about other's words, positionalities, and subjectivities without judgment. Then, we further reflect on how we understand their expressions, for language is all about interpretations, and we usually tend to interpret through our lived experiences. With internal dialogues, we can form a better sense through others' lived experiences, putting ourselves in their shoes, trying to understand their perspectives, even if they may be contradictory to our experiences or perspectives. We are doing so because we care about ourselves and others, trying to create a safe and respectful environment to allow different perspectives, voices, and narratives to happen naturally.

For the sake of illustration, we will present three micro-level processes of dialogue in evaluation and then exemplify each with a story from our evaluation practice to illustrate the purpose and role of internal dialogue. Even still, we view them as intertwined and occurring simultaneously. We also acknowledge micro-level processes intersect with the macro, social-political contexts. Finally, we do not view what we present as a complete understanding of micro-level processes for dialogue; instead, we are exemplifying some essential processes.

Dialogue as Reflection, Self-Care, Care for Others, and Building Trust

We recognize that people, including ourselves, enter evaluation spaces as individuals with positions based on intersecting identities. These intersecting identities (e.g., racial, gender, religious, political, and socioeconomic) are triggered and reflected by space dynamics, conversations, and the roles we play. In our stories, these triggering and reflective moments created feelings of discomfort, uncertainty, and sometimes disappointment. Holding and navigating these emotional moments required dialogue with self and others in the evaluation space. By engaging in dialogue with stakeholders or evaluation team members, we demonstrated care and built trust. These conversations were not always the easiest, but they were necessary for easing, not necessarily resolving, tension and moving the work forward. We noticed that when we experienced tension and discomfort in our evaluation work, we engaged in dialogue with others (e.g., family members, colleagues, friends) outside of the evaluand as a form of self-care. This processing exercise often felt necessary for emotional and mental regulation before returning to the evaluation space.

With a team of two staff and a partnering evaluation nonprofit organization, I[1] was working on an evaluation for a professional development program in a national network. Most staff attendees identified as White, yet they primarily served people of color. The program had a clear mother, the developer of the intervention, who was well respected and had worked for years to build the network. In the wake of public awakening to the racism pandemic in the United States, she wanted to implement a shared diversity, equity, and inclusion (DEI) vision in relation to their professional development at an annual national meeting. Our team had done some focus groups with trainers regarding how White supremacy culture was evident in their professional development. As an evaluator, I had been invited to be part of a committee that planned the event. A community rupture occurred at the national meeting with approximately 100 people present. Some culturally insensitive professionals caused harm to staff of color, and the staff of color expressed and named the harms. The community, including the planning committee and the mother, a White woman, wrestled with how to respond. I participated in discussions of the situation with the committee; shared concern for the harm done, as did others; used evidence from the evaluation to demonstrate the various experiences and perspectives of staff in the network; and interpreted this evidence using an established diversity framework in their field.

I listened to the concerns of stakeholders of color who had lost trust. I engaged in internal dialogue and wrestled with what I could have done better to intervene. Could I have done more to prevent the incident and thus the harm? Was this conflict good and necessary to repairing harm long term? I wondered if evaluation activities and privileging alternative voices may have provided some initial space to name inequities and welcome conflict prior to the public event. I realized that the staff of color did not need me to speak on their behalf.

Given my relationship with the mother (having mutual respect and shared affinity), I considered my role and responsibility to address the situation on behalf of my evaluation team, two of whom identified as people of color. I spoke with the mother directly and slowed down the evaluation work. I emphasized relationship building and efforts to decentralize leadership in the network and worked to ensure that evaluation activities worked within re-envisioned systems and interrogated them for perpetuating oppressive norms. With time and the addition of a DEI consultant that provided professional development across the network, I observed repair in relationships and changes in organizational structures that were perpetuating inequities. I noticed the mother and other members realized how much their work and network were embedded in White supremacy culture.

Dialogue as for Understanding Differences and Communicating Shared Values

We enter evaluation spaces and witness power differences and diversity of culture, ideas, and beliefs. These identity elements shape how people contribute and take space when a dialogue is taking place. As evaluators, we often felt the need to execute different roles, whether it was to be an observer, reflective listener, or facilitator, to make sure that stakeholders had equitable opportunities to speak. Sometimes stakeholders carried the dialogue on their own; therefore, the appropriate move was to simply observe and provide critical feedback afterward. On other occasions, we also experienced power differences within our own evaluation teams that forced us to reflect and discuss with colleagues what the power difference entailed and how it influenced the roles we played. Recognizing and holding space to dialogue about power imbalances prepared us to authentically engage with stakeholders as a collective front and allowed us to carry ourselves into the evaluation space, aware of our influence over group dynamics, dialogue, and decision-making. This particular awareness of power differentials also allowed us to be cognizant of how power was exercised and navigated. In our stories, dialogue brought tension, discomfort, and at times, misunderstanding. Engaging in dialogue challenged stakeholders, and even ourselves, to wrestle with values that guided the evaluation process and the evaluand. In a reflective session, one of us said,

> I think that through dialogue, we [as evaluators and stakeholders] can unearth the values that we're maybe operating from but do not always articulate well. We might say we value this thing but have conditions that are attached to it. Our actions and inactions are sometimes linked to those values.

From our experiences, we know that perspectives, beliefs, and values are not always articulated or translated accurately in speech and action. There is a Chinese poem, "不识庐山真面目，只缘身在此山中，" in translation, we are not fully aware of the true self on Mountain Lu because we are in the mountain. In other words, to appreciate a holistic picture, we need to step out of the situation and view that from different angles and perspectives. In evaluation, this disconnect is remedied by effective dialogue and legitimizes diversity within the evaluand, all of which are elements of democratization.

Being a novice evaluator, I was eager to provide recommendations based on my observations, communications, and interactions with different stakeholders of an after-school program for a nonprofit organization. Most student stakeholders would like to have fun activities rather than having their homework done with tutoring and support. Once they developed a friendship and a mentorship relationship with their close tutors, they preferred familiar tutors for longer time periods. Parent stakeholders prioritized having someone oversee their kids and tutor them, and they wanted to ensure that their kids finished their homework. Parents also preferred to spend less money. The organization stakeholders would like to make money, cut the budget, attract more parents to pay for their services, hire more volunteers to support their tutoring services, and find funding to support their operations. The organization did not prioritize the well-being or stability of their tutors. When I shared the findings and action recommendations with the organization stakeholders, I was surprised when the organization stakeholders acknowledged several issues. For instance, the after-school program had the potential to purchase more culturally relevant children's literature, equip with more technology devices, hire more experienced tutors and volunteers, provide training or professional development for newly hired tutors and volunteers, and maintain the stability of personnel.

After the dialogues within this group, I started to self-reflect on my roles and positionalities and make negotiations and pay more attention to the shared values among stakeholders. I wanted to improve the current program based on the shared values and the capacity of the organization's stakeholders. After conducting another round of internal dialogue, I realized that the organization stakeholders did well within their capacity to provide underserved students with snacks, drinks, tutoring services, indoor and outdoor activities, supervision, children's books, electronic devices, field trips, celebrations, and more. I decided to make a balance between being a critical evaluator and developing an asset mindset shift. So I made in-action recommendations based on the organization stakeholders' capacity, including provide incentives (such as a gift card) for volunteers, offer professional training for the hired volunteers, create a portfolio for volunteers to share their experiences and recommendations to better support the program, and work closely with universities to make their after-school program a community site for education major students to earn teaching and tutoring experiences. To achieve the long-term, I would recommend that the organization stakeholders look for funding and donors and launch campaigns to acquire resources and funding, such as collecting new or secondhand culturally responsive textbooks, tablets, and laptops.

Having dialogues with different stakeholders motivated me to form a holistic view and critical analysis in making negotiations between long-term goals and short-term goals within its capacity for improvement. Moreover, those dialogues within the team members and inner dialogues gave me the confidence to step out of my comfort zone, acknowledge and validate different perspectives, and shift my narrow, stereotyped ideas. Having inner dialogues supported me to understand the program from the perspectives of different stakeholders.

Dialogue as Resistance

Given the racial makeup of our group (Black, Asian, Pacific Islander, and White), we noticed, across our reflections, that we often used dialogue as resistance or self-preservation. We entered evaluation spaces hyper aware of racial identities and the power they assume. This reality dictated how we navigated, participated in, and used dialogue. As a resistance tool, we used dialogue to address tension or discomfort when stakeholders raised issues relevant to injustice or inequity. We also used it to challenge stakeholders to think about these issues and related values. Given our respective racial backgrounds, the degree to which we used dialogue to resist and challenge differed. This is a reflection of how racialized power influences how we choose to confront tensions and dialogue.

In some instances, speaking up and engaging was not always a decision we chose. The decision to remain silent or simply listen varied contextually. In some cases, choosing not to speak was a decision to balance power and resist existing, problematic norms related to power. In other cases, it was to provide more space for stakeholders to have a voice. Regardless, we recognized that balancing participation in dialogue to resist exacerbating problematic social norms is important for the democratization of the evaluation.

I worked for a non-American government entity as an internal researcher/ evaluator. One of my tasks was to evaluate a 4-year educational action plan with outcomes focused on student achievement, curriculum, and the school district's operations. To best capture the action plan's 2-year progress, I decided that a mixed-methods approach was most appropriate. In addition to quantitative data analysis of grades and assessments scores, I wanted to incorporate focus groups of various stakeholders and interview teachers. However, when I presented my evaluation design in a leadership meeting, I received pushback as some individuals questioned the appropriateness of including teachers in the process.

Within minutes, I had to wrestle with several things, internally, to decide on how to approach the situation. I was a native of the community that the school district served, but I was also an outsider in that I was not a part of the education system. The people in the room occupied higher professional and social positions than I did. They also had more experience working within the local education context. Furthermore, some of them were elders and members of my extended family. As direct users of the curriculum, I knew that the voice of teachers was essential for the evaluation and that they would be able to speak to its effectiveness in their schools. I ultimately decided to acknowledge the care and commitment to their joint work while advocating to interview teachers. I respectfully pushed back. Through my internal dialogue, I balanced how to move forward while respecting my culture, the organizational norms that existed, and the expectations of my job as an evaluator.

SEEDS OF CHANGE

Dialogue is a seed of change. It exists in different forms, with each form possessing catalytic potential. We are never certain about how it will grow or what it will yield, but we trust that when sown with intention and fostered with care, it blooms to inspire something new, unique, and unexpected. We found that our community of practice gave us essential support for using dialogue to promote a just society through our evaluation practice. In the midst of the COVID-19 pandemic—a difficult time in the world, our communities, and our lives—we formed an authentic community. We valued the trust and care that held us together, the recognition and sharing of power embedded with our identities, and the willingness to share stories for our collective learning. We grew and changed, as did our evaluation practice. Dialogue as a seed of change represents our transformative learning progress and our sincere hope for a just society. Our experience reaffirmed our conviction that dialogue in evaluation practice, although emotionally and mentally laborious at times, is justice work and therefore necessary

NOTE

1. We use "I" in each of our stories as a collective "We." Although an individual author in our community of practice originally told the story, we collectively chose which stories to represent in this chapter and contributed to the interpretation and representation of each story. Thus, an individual story became a collective story.

REFERENCES

Abma, T. A. (1997). Playing with/in plurality: Revitalizing realities and relationships in Rotterdam. *Evaluation, 3*(1), 25–48. https://doi.org/10.1177/13563 8909700300103

Abma, T. A. (1998). Text in an evaluative context: Writing for dialogue. *Evaluation, 4*(4), 434–454. https://doi.org/10.1177/135638909800400404

Boyce, A. (2019). A re-imagining of evaluation as social justice: A discussion of the Education Justice Project. *Critical Education, 10*(1), 1–17. https://doi.org/10.14288/ce.v10i1.186323

Fetterman, D. M., Kaftarian, S. J., Kaftarian, S. J., & Wandersman, A. (1996). *Empowerment evaluation: Knowledge and tools for self-assessment and accountability.* SAGE Publications.

Guajardo, A. D., Robles-Schrader, G. M., Aponte-Soto, L., & Neubauer, L. C. (2020). LatCrit theory as a framework for social justice evaluation: Considerations for evaluation and evaluators. *New Directions for Evaluation, 2020*(166), 65–75. https://doi.org/10.1002/ev.20409

Greene, J. C. (1997). Evaluation as advocacy. *Evaluation Practice, 18*(1), 25–35. https://doi.org/10.1016/S0886-1633(97)90005-2

Greene, J. C. (2000). Challenges in practicing deliberative democratic evaluation. *New Directions for Evaluation, 2000*(85), 13–25. https://doi.org/10.1002/ev.1158

Greene, J. C. (2001). Dialogue in evaluation: A relational perspective. *Evaluation, 7*(2), 181– 187. https://doi.org/10.1177/135638900100700203

Greene, J. C. (2005). Evaluators as stewards of the public good. In S. Hood, H. Frierson, R. Hopson (Eds.), *The role of culture and cultural context: A mandate for inclusion, the discovery of truth, and understanding in evaluative theory and practice* (pp. 7–20). Information Age Publishing.

Greene, J. C. (2006). Evaluation, democracy, and social change. In I. F. Shaw, J. C. Greene, & M. M. Mark (Eds.), *The SAGE handbook of evaluation* (pp. 118–140). SAGE Publications.

Greene, J. C. (2007). *Mixed Methods in Social Inquiry.* Jossey-Bass.

Greene, J. C. (2015). Culture and evaluation from a transcultural belvedere. In S. Hood, R. Hopson, & H. Frierson (Eds.), Continuing the journey to reposition culture and cultural context in evaluation theory and practice (pp. 91–107). Information Age Publishing.

Guba, E. G., & Lincoln, Y. S. (1989). *Fourth generation evaluation.* SAGE Publications.

Hall, J. N. (2020). The other side of inequality: Using standpoint theories to examine the privilege of the evaluation profession and individual evaluators. *American Journal of Evaluation, 41*(1), 20–33. https://doi.org/10.1177/1098 214019828485

Hood, S. (2000). Commentary on deliberative democratic evaluation. *New Directions for Evaluation, 2000*(85), 77–83. https://doi.org/10.1002/ev.1163

Hood, S. (2001). Nobody knows my name: In praise of African American evaluators who were responsive. *New Directions for Evaluation, 2001*(92), 31–44. https://doi.org/10.1002/ev.33

Hopson, R. K., Kirkhart, K. E., & Bledsoe, K. L. (2012). Decolonizing evaluation in a developing world: Implications and cautions for equity-focused evaluations. In M. Segone (Ed.), *Evaluation for equitable development results* (pp. 59–83). UNICEF.

House, E. R., & Howe, K. R. (1998). The issue of advocacy in evaluations. *American Journal of Evaluation, 19*(2), 233–236. https://doi.org/10.1016/S1098 -2140(99)80200-8

House, E. R., & Howe, K. R. (1999). *Values in evaluation and social research.* SAGE Publications.

Johnson, J., Hall, J., Greene, J. C., & Ahn, J. (2013). Exploring alternative approaches for presenting evaluation results. *American Journal of Evaluation, 34*(4), 486–503. https://doi.org/10.1177/1098214013492995

Lemos, D., & Garcia, D. (2020). Promoting culturally responsive and equitable evaluation with Latinx immigrants. *New Directions for Evaluation, 2020*(166), 89–100. https://doi.org/10.1002/ev.20410

Mathison, S. (1996). Evaluation as a democratizing force in schools. *International Journal of Social Education, 11*(1), 40–47.

Mathison, S. (2000). Deliberation, evaluation, and democracy. *New Directions for Evaluation, 2000*(85), 85–89. https://doi.org/10.1002/ev.1164

Mcbride, D. (2015). Cultural reactivity vs. cultural responsiveness: Addressing macro issues starting with micro changes in evaluation. In S. Hood, R. Hopson, & H. Frierson (Eds.), *Continuing the journey to reposition culture and cultural context in evaluation theory and practice* (pp. 179–202). Information Age Publishing.

Mcbride, D., Casillas, W., & LoPiccolo, J. (2020). Inciting social change through evaluation. *New Directions for Evaluation, 2020*(166), 119–127. https://doi.org/10.1002/ev.20405

Merriam, S. B., & Bierema, L. L. (2014). *Adult learning: Linking theory and practice.* Jossey-Bass.

Mertens, D. M. (1999). Inclusive evaluation: Implications of transformative theory for evaluation. *American Journal of Evaluation, 20*(1), 1–14. https://doi.org/10.1177/109821409902000102

Mezirow, J. (1991). *Transformative dimensions of adult learning.* Jossey-Bass.

Mezirow, J. (1997). Transformative learning: Theory to practice. *New Directions for Adult and Continuing Education, 1997*(74), 5–12. https://doi.org/10.1002/ace.7401

Pine, G. J. (2009). *Teacher action research: Building knowledge democracies.* SAGE Publications.

Rallis, S. F., & Rossman, G. B. (2000). Dialogue for learning: Evaluator as critical friend. *New Directions for Evaluation, 2000*(86), 81–92. https://doi.org/10.1002/ev.1174

Reid, A. M., Boyce, A. S., Adetogun, A., Moller, J. R., & Avent, C. (2020). If not us, then who? Evaluators of color and social change. *New Directions for Evaluation, 2020*(166), 23–36. https://doi.org/10.1002/ev.20407

Ryan, K., Greene, J., Lincoln, Y., Mathison, S., Mertens, D. M., & Ryan, K. (1998). Advantages and challenges of using inclusive evaluation approaches in evaluation practice. *American Journal of Evaluation, 19*(1), 101–122. https://doi.org/10.1177/109821409801900111

Schwandt, T. A. (1997). Evaluation as practical hermeneutics. *Evaluation, 3*(1), 69–83. https://doi.org/10.1177/135638909700300105

Stake, R. E. (2000). A modest commitment to the promotion of democracy. *New Directions for Evaluation, 2000*(85), 97–106. https://doi.org/10.1002/ev.1166

Wenger, E. (1998). *Communities of practice: Learning, meaning, and identity.* Cambridge University Press.

Wenger, E. (2000). Communities of practice and social learning systems. *Organization, 7*(2), 225–246. https://doi.org/10.1177/135050840072002

Wenger, E. (2010). Communities of practice and social learning systems: The career of a concept. In C. Blackmore (Ed.), *Social learning systems and communities of practice* (pp. 179–198). Springer.

CHAPTER 14

EXPLORING THE LINGUISTIC IMPLICATIONS OF DIALOGUE IN EVALUATION

Jill Anne Chouinard
University of Victoria

ABSTRACT

Language is a key part of all social interactions, as it shapes how people understand and construct meaning, make sense of the world, coproduce knowledge, and build a common identity as members of a community of practice. In this chapter, I reframe dialogue (and its sociopolitical dimensions) as a matter of language, power, and voice, all of which I envision shaping and constraining the potential for meaningful and equitable dialogue. This reframing looks at linguistic concerns with dialogue from a micro and macro perspective, framing these multitiered linguistic influences as dialogically significant in diverse evaluation contexts. When we look across at the micro and macro implications of language on our dialogic processes in evaluation, we can see that language matters, as it not only represents who speaks, who remains silent, whose words, concepts, and ideas are given voice, and whose remain hidden, but also whose concepts, meanings and interpretations are given final script.

Disrupting Program Evaluation and Mixed Methods Research for a More Just Society, pages 199–211

I am in dialogue with program beneficiaries

with sex offenders who have spent decades in prison for raping their children

with teachers whose students are considered untouchable, unworthy, even dirty

with women living in remote, fly-in communities, unable to escape violence
    until the Spring thaw

I am in dialogue with evaluands

with youth who have witnessed collective suicide pacts and been spared

with girls whose narratives of sexual assault are long lost in government archives

with grandmothers caring for grandbabies as they try to understand fentanyl

I am in dialogue with stakeholders

with midwives who tell stories about childbirth in remote Northern communities

with Hispanic youth who fear waking up without their parents

with community organizers who struggle to protect young children from gang
    violence

I am in dialogue with the subject

with LGBTQ youth whose own parents chill at their sight

with teachers whose students come to school with bruises and lost looks

with young mothers whose babies are born prematurely, addicted to crack

I am in dialogue with people

whose vulnerability shapes stories of injustice, despair, even hope

I listen

(inwardly I rage at injustice, I cry),

as the evaluator turns on the audiotape, takes notes, asks questions

reframes, retheorizes, and analyzes other peoples' stories

into narratives that remain so seemingly,

so very far away

I am in dialogue with myself

As I question whose words, whose concepts, and whose fear really fill the pages

This poem represents my attempt to capture the epistemological and ontological unease I feel about the concept of dialogue given the diverse, complex, and messy contexts that characterize so much of our work as evaluators. In academia, many of us lead privileged lives of comfort, sharing few
concerns that beset much of the world's population. Our methodological

toolkit, which is our approaches, frameworks, language, methods, and learned processes, helps us maintain a level of academic and professional reserve. Our words, concepts, and theories provide a methodological scaffolding to support our practice's dialogical and relational contours. In the current historical moment, which can be defined by stark disparities in power, privilege, and voice, learning *how* to navigate these complex sociocultural contexts to address issues of inequity remains paramount. Equally daunting, and much less theorized outside of culturally responsive evaluation contexts, is what we as evaluators bring to the table, and how who we are and what we bring profoundly shapes the dialogic process. As Val Napoleon (2004) asks, "Who gets to say what happened?" Whose authorship counts? In her privileging of concepts such as relationships, dialogue, values, advocacy, inclusion, diversity, and voice, Jennifer Greene reframes the possibility of praxis and, in so doing, shifts us from a sense of lingering, sense of despair to a fragile and meaningful engagement with hope.

In this chapter, I extend Jennifer Greene's conception of dialogue in evaluation to explore the micro and macro implications of language in evaluation contexts where concern with diversity, power, and privilege prevails. At issue is not dialogue as a social, normative, or even political ideal, but rather how to uphold such an ideal in evaluative contexts fraught with sociocultural and political inequities. As Greene (2001) herself notes, despite the genuinely democratizing potential of dialogue in evaluation, engaging in dialogue is not without its challenges, as it opens up what she refers to as the "socio-political dimensions of communication" (p. 183). Thus, while not rejecting the democratic potential (or necessity) of dialogue in evaluation, I frame dialogue (and its sociopolitical dimensions) as matters of language, power, and voice, all of which I envision as shaping and constraining the potential for meaningful and equitable dialogue.

In the first part of this chapter, I describe Jennifer's conception of dialogue and outline some of the challenges she has experienced advocating for dialogue in collaborative contexts where participants have diverse perspectives, experiences, values, and beliefs. The chapter then sketches the boundaries of dialogue in evaluation, extending Jennifer's conception to include the linguistic aspects of dialogue which, as my poem suggests, has profound implications for how we make meaning in co-creating knowledge with others in diverse evaluation contexts. I then shift to the linguistic concerns with dialogue from both a micro and macro perspective, framing these multitiered linguistic influences as dialogically significant in diverse evaluation contexts. I first begin with how I came to know Jennifer.

PREAMBLE

My first encounter with Jennifer was through her scholarship. During my first evaluation course, we were required to select an evaluation scholar for a poster presentation. With little knowledge about evaluation or about the diversity of the field or range of perspectives, I glanced apprehensively through the list of unfamiliar names that our professor had preselected. I had a single criterion—my scholar had to be a woman, and, with few choices available, I selected Jennifer. Happy with my choice, I began reading her articles in earnest, and there were many. The professor had casually mentioned that we might consider contacting our selected evaluation scholar by email for a possible interview. Reading through her articles, I was completely absorbed by Jennifer's clear and insightful prose, by the strength of her values, all of which I have come to know is reflected in her writing, her work as an evaluator, and in her scholarship. By the time I sent Jennifer an email, I had developed my first (and enduring) academic crush.

My political and social activism had always led me to work in politics, join picket lines, and march in demonstrations, but in reading Jennifer's articles, I realized that epistemology, how we create knowledge as social scientists and evaluators, is exceedingly political. Those we engage with and exclude in an evaluation, the standards we use, and the defined purposes of an evaluation are all fundamentally political decisions that we make as evaluators, reflecting our values and our broader sociopolitical and cultural aspirations. I completed my master's in education, and, inspired by Jennifer's ideas about the potential for evaluation to engage with issues of inequity and democratizing social inquiry practices, I shifted my PhD focus to program evaluation. I now consider myself a scholar activist, a weaver of histories, traditions, stories, and voices, as I always think about borders and boundaries, who is in or out, and whose voices are privileged and whose are silenced in my work as an evaluator, teacher, and scholar. Thus, my discussion of dialogue reflects my underlying fear that as evaluators (and academics), the potential of our impact will always remain limited, and at best, partial. I wonder aloud what holds us back in having a greater impact, what shapes and influences our dialogue with others, and how we can use our evaluation practice (and scholarship) in service to creating a more equitable and just world. I feel strongly that we engage in these ideas within the context of evaluation because Jennifer (and like-minded people) pushed the boundaries of our field away from its more modernist roots, toward a vision of evaluation that aspires to create a more equitable and democratic future.

JENNIFER'S CONCEPTION OF DIALOGUE

In many important respects, genuine dialogue is meaningful evaluation.
—Greene, 2001a, p. 184

Dialogue lies at the heart of Jennifer Greene's contributions to social inquiry. Whether in terms of relationships, method choice, values, or advocacy, her commitment to dialogue valorizes principles of inclusion, engagement, diversity, and democracy. In evaluation, dialogue also holds promise for democratizing practice through the inclusion of multiple perspectives and voices and engagement with diverse others and, as such, is given priority through participatory, transformative, and culturally responsive approaches. In diverse community contexts, relationships between evaluators and stakeholders are mediated across cultural, class, gender, and racial divides and across political, economic, and social histories often defined by unequal status, power, privilege, and voice. Despite the challenge, in many ways, the dialogic process of coming to know self and others is fundamental to our understanding of what it means to build relationships in culturally diverse program and evaluation settings.

For Greene (2000, 2001a, 2001b), dialogue in evaluation is conceptualized as more than a simple transactional, extractive, or casual exchange of information, as it engages participants intentionally with notions of inclusivity, equity, diversity, and respect, all foundationally democratic concepts. As Greene (2001a) explains,

> It is through the defining lens of value commitments, rather than methodology or purpose, that dialogue in evaluation is most meaningfully understood and discussed. For dialogue in evaluation most fundamentally means a value commitment to engagement, engagement with problems of practice, with the challenges of difference and diversity in practices and their understandings, and thus with the relational, moral and political dimensions of our contexts and our craft. (p. 181)

For Greene, dialogue directly addresses the challenges of diversity, difference, and power locally within the evaluative context. At the same time, dialogue also speaks to her aspirations for creating a more democratic and equitable society. These aspirations echo beyond the evaluation program's context, as the gap between wealthy and poor, marginalized and privileged, and racialized and White, grows daily, both here at home and across the globe.

Whether discussing participatory evaluation, performance measures, mixed methods, values-based evaluation, or advocacy, for Greene, the concept of dialogue speaks to her moral and ethical commitment to the values of equity, inclusion, and democracy (Greene, 2001a), to her notion of "evaluator as engaged person" (King & Stevahn, 2002, p. 3), and to her vision of quality in evaluation (Greene, 2001b). From an emic perspective, dialogue opens up space for all legitimate participants to interact, communicate, co-learn, and share views and perspectives, all of which directly engage with the moral, ethical, and political implications of the dialogic process of social inquiry. For Greene, dialogue represents a democratic commitment to ensuring that everyone has a seat at the table, that no legitimate voices are excluded from the conversation.

While dialogue is essential at the local level for engaging participants in relationships of trust and respect, it is also seen as a way to address issues of diversity and power in the community. In collaborative contexts, dialogue is seen as a key feature of how evaluative claims are made to judge program quality, as well as a constitutive characteristic of the evaluative claim itself (Greene, 2001). Monologic processes, those exemplified by the absence of perspective and voice, are considered by Greene (2001b) "neither complete nor warranted" (p. 66). Judgments of quality thus hinge on dialogue, relationship, inclusion, voice, and diversity of perspective. Participatory evaluators thus share a responsibility for ensuring that dialogic and relational values and commitments are honored, supported, and upheld (Greene, 1997).

For Greene, the notion of dialogue extends beyond local engagement, inclusivity, shared perspectives, and voice, and beyond data quality to a more equitable and democratic vision and aspiration. As Greene (2001a) describes, "Dialogic evaluation is about not just interpersonal; but also political equity of voice, understanding and acceptance of difference, and re-distribution of resources" (p. 185). The range and depth of Greene's commitment to dialogue ultimately speak to her vision about the potential for evaluation to play a role far beyond its technical attributes, to one that challenges the status quo by providing a dialogic space for the inclusion of a diversity of perspectives, voices, and values. As Greene (2001a) notes, "Dialogic evaluation then serves as a forum and force for democratization" (p. 185).

Despite these aspirations, as Greene has experienced (see Greene, 2000, 2001; Mathie & Greene, 1997), evaluation can magnify issues of power and politics, especially in socially and culturally diverse contexts. In their attempt to create democratizing conversations in a diverse evaluation context, Mathie and Greene (1997) experienced tension between representation and engagement, which led them to question whether too much diversity might impede the transformative potential of collaboration. Describing another evaluation experience, Greene (2000) noted that despite her attempt to build an inclusive and dialogic process for the evaluation of a local high school science curricula, the power of local elites circumvented her dialogic and democratic vision for the evaluation. While the effect of local power was evident, its impact circumscribed the diversity of perspectives at the table and shaped how evaluation was conceptualized and ultimately received. As Greene (2000) explains, the evaluation of the high school science curriculum "was neither inclusive nor dialogic, neither deliberative nor democratic, despite my repeated attempts to make it so" (p. 16).

In another example, Greene (2001b) describes the challenges she experienced creating a dialogic, democratic, and equitable space with a community coalition during the evaluation of discriminatory workplace practices. As Greene (2000) describes, "While this participatory evaluation intentionally intended to broaden and democratize ownership and decision authority over the project, the actual pattern of participation and power served primarily to reinforce existing power inequities. Discourse gave way to disengagement, engagement to distrust" (p. 67). While daunting, these experiences serve as a stark reminder that dialogue, especially among a diverse group of people, is never straightforward. In what follows, I reconsider the local and broader level "sociopolitical dimensions of communication." I extend the concept of dialogue to include its linguistic attributes, as I consider language a key part of all social interactions, as it shapes how people understand and construct meaning, make sense of the world, gain control over the work they perform (e.g., coproducing knowledge), and build a common identity (e.g., as members of a community of practice).

THE LINGUISTIC DIMENSIONS OF DIALOGUE

As the foregoing makes clear, the "culture of power" (Delpit, 1988) that characterizes and threatens our society's sociopolitical and cultural fabric is difficult to shake. Spivak's (1988) question, "Can the subaltern speak?" (p. 24), points us to several key concerns: Who frames the dialogue? Who decides who speaks and who does not? Whose words and concepts shape meaning? Which meaning systems define the rules of engagement? Who narrates these terms? Who is included or excluded? Who is silenced? Who shapes the conversation? Who benefits from this dialogue? Who is at risk? While challenging, these questions shift our focus away from the notion of evaluation as a set of relations and engagements fixed in time and space to one that situates evaluation as a set of relations connected to a larger sociocultural and political system that influences and shapes the local setting. These influences are significant; they shape the layers, contours, and dynamics of the dialogic process, as evaluators and participants, each from their own unique sociocultural backgrounds (and often from quite distinct linguistic cultures), together co-construct evaluative knowledge. As Burbules (2000) reminds us,

> Dialogue is not simply a momentary engagement between two or more people; it is a discursive relation situated against the background of previous rela-

tions... and [it] may impinge... in ways that may shape or limit the possibilities of communication and understanding. (p. 263)

Our words, language, and dialogue have materiality (Burbules, 2000; Soja, 1989); they come from somewhere, have a history and a sociocultural and political context that influences what is said and how it is said and by whom. Within the context of dialogically engaged evaluation practice, language is thus an especially important consideration when Western-based researchers work in culturally, socially, and linguistically diverse community settings. In what follows, I look at the micro and macro dimensions to highlight the profound influences of language in the dialogic encounter among participants and evaluators.

Micro Influences

Locally, dialogue is shaped by the cultural context of the community and its social, political, and economic history (especially its history with the dominant society), all of which influences interpersonal relationships between evaluators and participants. There are also differences in language and communication styles, levels of literacy in the community, and cultural differences, all factors that influence the context and the dialogical space. While there may be cultural and linguistic congruity between evaluators and participants, there is nonetheless what Dahler-Larsen et al. (2017) refer to as the "evaluator language" (p. 116), a codified and privileged set of practices, approaches, discourses, and rules that are often at odds with "local languages" (p. 116). The issue here is not simply about surface differences in communication style and approach, but about whose language and whose words frame the dialogue and whose linguistic ground rules are given priority. This challenge, perhaps expressed locally as a challenge of "voice," also raises issues of "hermeneutical marginalization" (Fricker, 2007, p. 147), a form of injustice where people lack the language skills, words, or knowledge to put their own experiences into words. Unable to participate in the conversation and being without the words or language to describe experiences puts participants at a serious epistemic disadvantage (Frank, 2013), an especially significant concern in collaborative practice. Whether people choose to participate or not, they can only participate if they are familiar with the rules of the game.

Some thus question whether such disparities of voice can be addressed or overcome through dialogue (de Castell, 2004), as it may serve as a barrier to genuine conversation (Jones, 2004). Others question whether our notion of dialogue in its idealized form is untenable in the current moment, especially given that disparities of class, race, gender, and power remain so profound (Ellsworth, 1989). Still, others argue that all voices are not equal, as history does not afford them equal epistemic weight (Boler, 2004; Macdonnell, 1986), an ongoing issue that puts participants at a further disadvantage as they may be unwilling to share their views or perspectives with others for fear that their "testimonies" will be misconstrued or ignored (Fricker (2007). These linguistic concerns are particularly relevant when evaluators enter into and work in evaluation contexts in which they lack cultural, social, or linguistic fluency. Also included at the micro level, local communication styles reflect local culturally and contextually distinct norms and values. These rules of protocol are characterized by who can speak, in what order, and who can speak for whom. Varying levels of literacy among participants also help define micro-level dialogic interactions, as evaluators must adapt and adopt evaluation approaches to meet the varying literacy needs of all participants (e.g., pictures, storytelling, poetry, art).

Macro Influences

A broader, more macro-level perspective embeds the local dialogic context within a larger, fundamentally interconnected, and dynamic sociocultural, political, and historical system that influences and shapes the parameters of local dialogue. It represents the powerful political, cultural, and historical dominance of Western inquiry practices (encapsulated in the Western canon), what for Foucault (1972) is a "cultural archive." This archive reflects the West's storehouse of organized, classified, and arranged knowledge, traditions, and stories (Smith, 1999). While it is taken almost completely for granted, we use it to reflect and look at ourselves. This archive is not neutral, as it excludes the positions, perspectives, and worldviews of the many peoples whose cultural and ethnic histories remain outside of the dominant Euro-Western, White, male view. The discourses of the social sciences (and of evaluation), our "ways of knowing," are well represented in this archive, as it contains our language, frameworks, processes, and standards of practice, all of which we use to shape, frame, and define the words and concepts for describing, representing and understanding the evaluand. While collaborative practices are designed to engage participants in a dialogic process so that they can be a central part of the evaluation process, our methodologies and evaluation approaches nonetheless remain affixed to modernist, Western ideals. The active inclusion of participants in our practice, while essential, cannot obscure evaluation's cultural authority or its power to define academic discourse, what Reagan (1996) might refer to as "epistemological ethnocentrism" (p. 5), where the parameters of what is considered "legitimate discourse" is delimited, predefined, and fixed.

DISCUSSION AND CONCLUDING THOUGHTS

I believe that coming to terms with human diversity and difference is the most important global challenge we currently face.
—Greene, 2001a, p. 186

The notion of authentic dialogue in this historical sociopolitical and cultural moment remains an ongoing and persistent challenge that I argue requires a critical reevaluation of dialogue to include its linguist characteristics to reposition it as a cultural, political, historical, and social construct. As evaluators and participants often communicate in different spoken languages and discourses within the same language, a linguistic focus brings awareness to issues related to translation, interpretation, representation, meaning-making, power, politics, and authorship. *Whose words, whose concepts, and whose fear really fill the pages?* When we look across at the micro and macro implications of language on our dialogic processes in evaluation, we can see that language matters. It not only represents who speaks, who remains silent, whose words, concepts, and ideas are given a voice, and whose remain hidden, but also whose concepts, meanings, and interpretations are given final script. Despite our aspirations, methodology—even collaborative or culturally responsive approaches—cannot assure meaningful dialogic proximity or authentic communication with participants (Chouinard & Cram, 2020). It is we evaluators who *reframe, retheorize, and analyze other peoples' stories* and *our* words and concepts that provide the final textual or numeric representation.

Understanding, listening, and engaging in dialogue across differences requires radical critique, an appreciation for dialogue as resistance (Everitt, 1996). As Greene has noted (see Greene, 2001a, 2001b; Matthie & Greene, 1997), it is not enough to invite everyone to the table, as power, voice, authority, status, and authorship cannot be so easily settled, especially in contexts defined by diversity of perspective, culture, power, and privilege. We need to acknowledge who has the power to frame and shape the dialogue, the words, concepts, and meanings, and author the final text, what Delpit (1988) refers to as the "culture of power" (p. 280). As Boler (2004) has argued, "All speech is not free...all voices do not carry the same weight...and [citing Butler she adds] social and political culture predetermines certain voices and articulations as unrecognizable, illegitimate, unspeakable" (p. 3). Dialogue remains an explicit political encounter with the other (Jones, 2004). From this perspective, dialogue cannot be considered a means to an end, a mere technique, but must be located as a web of relations embedded in a living cultural and political history; language has materiality (Bakhtin, 1981; Ellsworth, 1989). As Kohn (2000) points out, language is not a transparent medium, and communication is not neutral, but even in democratic contexts, it can mask the hierarchical distribution of linguistic competence and reproduce dominant exclusions. The issue is not with our intentions but with the nature and constitution of language itself. Wittgenstein's (1922) well-known quote comes to mind here, "The limits of my language mean the limits of my world" (Sec. 5.6). We can only take part in dialogue if our familiarity extends to an understanding of the social, cultural, and political conditions of the text. There is a precondition to participation in dialogue, which makes democratizing dialogic practice in evaluation, especially if we or our participants do not understand each other's "language games." Thus, while dialogue does provide the opportunity for what Everitt (1996) calls "practice talk," a concept she used to describe the democratic opportunities dialogue offers to cocreate alternative discourses, it does not present itself unfettered by history.

As evaluators, language is often only highlighted as a concern when we are working in linguistically diverse settings and when we are unfamiliar with the spoken language. Yet language is a part of how we communicate with one another, and its significance extends far beyond the transactional play of words, the formal and informal styles of communication structures and styles, and even the levels of literacy. As Shor and Freire (1987) describe, "Communicating is not mere verbalism, not a mere ping pong of words and gestures. It affirms or challenges the relationship between the people communicating, the object they are relating around, the society they are in" (pp. 13–14). As evaluators, how do we accommodate different local languages, communication styles, and language literacy? Far too often, linguistic issues in evaluation are primarily framed as technical, often reduced to hiring translators or cultural brokers from the local community to help us make sense of cultural and linguistic differences (see Chouinard & Cram, 2020). Yet, we know from experience that neither our methods nor our "expertise" can shield us from the dialogic challenges we encounter in our evaluation practice. However, it is only in opening up these dialogic spaces that we can create (and cocreate) a dynamic and fluid world, crowded by a diversity of voices, perspectives, and values and by uncertainty and possibility.

It is our responsibility as evaluators to attend critically to the "cultural politics of dialogue" (Burbules, 2006, p. 112), and to acknowledge that the power we have as evaluators is ultimately the power to engage across differences and to explore the democratic potential of dialogic approaches to evaluation. As Greene (2001a) points out, engaging dialogically across differences can lead to mutual learning (about ourselves, others, and our identities), legitimize a diversity of interests, and, in the longer term, help us build toward reconciliation. At the same time, I believe we have a responsibility to acknowledge that language, as a cultural, political, and historical construct, is a key part of dialogue, as it shapes how people understand and construct meaning and how they make sense of their world. In a world so defined by stories of racial, social, and economic inequity, injustices, and environmental disaster, it is a responsibility worth taking seriously.

REFERENCES

Bakhtin, M. (1981). *The dialogic imagination: Four essays by M. M. Bakhtin.* University of Texas Press.

Boler, M. (2004). All speech is not free: The ethics of "affirmative action pedagogy." In M. Boler (Ed.), *Democratic dialogue in education: Troubling speech, disturbing silence* (pp. 3–13). Peter Lang Publishing.

Burbules, N. C. (2000). The limits of dialogue as a critical pedagogy. In P. Trifonas (Ed.), *Revolutionary pedagogies: Cultural politics, education, and discourse theory* (pp. 251–273). Routledge.

Burbules, N. C. (2006). Rethinking dialogue in networked spaces. *Cultural Studies↔Critical Methodologies, 6*(1), 107–122. https://doi.org/10.1177/1532708 605282817

Chouinard, J. A., & Cram, F. (2020). *Culturally responsive approaches to evaluation: Empirical implications for theory and practice.* Sage.

Dahler-Larsen, P., Abma, T., Bustelo, M., Irimia, R., Kosunen, S., Kravchuk, I., Mina, E., Segerholm, C., Shiroma, E., Stame, N., & Tshali, C. K. (2017). Evaluation, language, and untranslatables. *American Journal of Evaluation, 38*(1), 114–125. https://doi.org/10.1177/1098214016678682

de Castell, S. (2004). No speech is free: Affirmative action and the politics of give and take. In M. Boler (Ed.), *Democratic dialogue in education: Troubling speech, disturbing silence* (pp. 51–56). Peter Lang Publishing.

Delpit, L. D. (1988). The silenced dialogue: Power and pedagogy in educating other people's children. *Harvard Educational Review, 58*(3), 280–298. https://doi.org/10.17763/haer.58.3.c43481778r528qw4

Ellsworth, E. (1989). Why doesn't this feel empowering? Working through the repressive myths of critical pedagogy. *Harvard Educational Review, 59*(3), 297–325. https://doi.org/10.17763/haer.59.3.058342114k266250

Everitt, A. (1996). Developing critical evaluation. *Evaluation, 2*(2), 173–188.

Foucault, M. (1972). *The Archaeology of knowledge and the discourse on language.* Pantheon Books.

Frank, J. (2013). Mitigating against epistemic injustice in educational research. *Educational Researcher, 42*(7), 363–370.

Fricker, M. (2007). *Epistemic injustice: Power and the ethics of knowing.* Oxford University Press.

Greene, J. C. (1997). Evaluation as advocacy. *American Journal of Evaluation, 18*(1), 25–35.

Greene, J. C. (2000). Challenges in practicing deliberative democratic evaluation. *New Directions for Evaluation, 2000*(85), 13–26.

Greene, J. C. (2001a). Dialogue in evaluation: A relational perspective. *Evaluation, 7*(2), 181–187. https://doi.org/10.1177/135638900100700203

Greene, J. C. (2001b). The relational and dialogic dimensions of program quality. In A. P. Benson, D. M. Hinn, & C. Lloyd (Eds.), *Vision of quality: How evaluators define, understand and represent program quality* (pp. 57–71). Emerald Group Publishing Ltd.

Jones, A. (2004). Talking cure: The desire for dialogue. In M. Boler (Ed.), *Democratic dialogue in education: Troubling speech, disturbing silence* (pp. 57–68). Peter Lang Publishing.

Kohn, M. (2000). Language, power, and persuasion: Towards a critique of deliberative democracy. *Constitutions. 7*(3), 408–429. https://doi.org/10.1111/1467-8675.00197

Macdonnell, D. (1986). *Theories of discourse.* Basil Blackwell.

Mathie, A., & Greene, J. C. (1997). Stakeholder participation in evaluation: How important is diversity? *Evaluation & Program Planning, 20*(3), 279–285. https://doi.org/10.1016/S0149-7189(97)00006-2

Napoleon, V. (2004). Who gets to say what happened? Reconciliation issues for the Gitxsan. In C. Bell & D. Kahane (Eds.), *Intercultural dispute resolution in Aboriginal contexts* (pp. 176–195). University of British Columbia Press.

Reagan, T. (1996). *Non-western educational traditions: Alternative approaches to educational thought and practice.* Routledge.

Shor, I., & Freire, P. (1987). What is the "dialogical method" of teaching? *The Journal of Education, 169*(3), 11–31.

Smith, L. T. (1999). *Decolonizing methodologies: Research and indigenous peoples.* Zed Books.

Soja, E. W. (1989). *Postmodern geographies: The reassertion of space in critical social theory.* Verso.

Spivak, G. C. (1988). Can the subaltern speak? In C. Nelson & L. Grossberg (Eds.), *Marxism and the interpretation of culture.* Macmillan.

Wittgenstein, L. (1922). *Tractatus-logigo-philosphicus.* Benediction Classic.

CHAPTER 15

HOLDING SPACE FOR EXPLORATION, REFLECTION, AND GROWTH[1]

Sarita Davis
Georgia State University

Alexis Kaminsky
Kaminsky Consulting, LLC

ABSTRACT

In a dialogue of former students who have been profoundly influenced by the scholarship and mentor experiences of Jennifer Greene, Davis and Kaminsky discuss Jennifer's influence on how they view the role of valuing in evaluation and how they incorporate valuing in their own work in the contexts of social work, Africana studies, education, and community evaluative practices. While reflecting on their introduction to the concept of valuing as graduate students of Greene at Cornell University, they weave scholarship, lived experiences, and core aspects of valuing, inclusivity, and social justice into their contemporary practices and approaches. In doing so, they emphasize the role of learning and dialogue in nonconventional, nondominant spaces that privilege new realities and epistemologies where we evaluators "hold space for each other."

Disrupting Program Evaluation and Mixed Methods Research for a More Just Society, pages 213–223
Copyright © 2023 by Information Age Publishing
www.infoagepub.com
All rights of reproduction in any form reserved.

> *Evaluation is . . . intrinsically judgmental, involving some criteria of "goodness"*
> *upon which judgements of quality and effectiveness are made. Evaluation is thus also*
> *intrinsically infused with values, because the selection of these criteria—alongside*
> *other evaluative decisions—rest on the privileging of some set of values over others.*
>
> —Greene, 2011, p. 193

Little did we know that such a statement was so provocative to so many evaluators. When we joined Cornell in 1991, debates raged about quantitative vs. qualitative (mixed methods came later), post-positivism vs. constructivism vs. critical theory, advocacy vs. objectivity, and whose knowledge mattered. Jennifer helped us understand how these conversations around methods, the nature of knowledge, evaluator role, and data, at their core, were about values. The humanistic, engaged character of evaluation that we learned from Jennifer at Cornell resonated with us and from where we each were coming to evaluation.

Sarita came to the field of program evaluation after working 2 years as a social worker in South Central Los Angeles. She quickly became frustrated with the imposition of Eurocentric ideas and values in determining "desirable" program outcomes in the lives of Black and brown program participants. Whose truth was being privileged? Where were the voices of the people? Under Jennifer's mentoring, Sarita found a kindred spirit with the heart of an abolitionist. She was already blazing a path of methodological inclusivity that resonated with Sarita's desire to find self-determination and liberation in programs targeting marginalized social service recipients.

Alexis came to program evaluation after spending time in Kenya, working at a community center, and exploring issues of power and multicultural education. In Jennifer, Alexis found a mentor and scholar who emphasized the *value* part of evaluation and encouraged her students to consider critically the ways in which evaluation practice maintained or challenged existing relationships. Jennifer's scholarship, particularly around the role of the evaluator, attention to pluralistic voices, and evaluation's educative potential, brought more heady topics of ethics, values, and social change to a practical point that has become the basis of Alexis's own evaluation practice.

One value that spoke to Alexis and Sarita most insistently was the inclusion of people whose perspectives were often marginalized in evaluation. Sarita, as Bill Trochim's student, initially turned to concept mapping as a means to engage stakeholders in evaluation practices, and with Bill and Jennifer's encouragement, became an early leader in diversifying evaluation's White, western complexion. Alexis, as Jennifer's student, focused on the potential of participatory and collaborative approaches to evaluation to make the process of evaluation more democratic and the outcomes more responsive to a variety of stakeholder interests.

In the ensuing 25-plus years, Alexis and Sarita have practiced evaluation in many contexts and applied a variety of approaches to bring in perspectives often underappreciated in evaluation. In the dialog that follows, they talk about how values show up in their own work and how Jennifer influenced their own efforts to make space for others' exploration, reflection, and growth in and through evaluation practice.

THE ROLE OF VALUING IN EVALUATION

SD: A central tenant of Greene's work is acknowledging that values are opaque—but powerful actors—on the stage of evaluation. As such, it is important to understand what constitutes valuing in evaluation. The explicit call to engage our values resonates with me as a social worker and as a professor in Africana studies because both disciplines encourage transparency of inherent values. The core values of the National Association of Social Work are service, social justice, dignity, and value for human relations. Africana studies is interdisciplinary and grounded in social justice. For me, inclusion doesn't happen without a commitment to action. I can be freely intentional about my commitment to transforming a problem or, as I look at it, an opportunity. The role of values in evaluation is central to the work of evaluation because we sit in judgment.

AK: Your comments make me think of the article Greene wrote with Hall and Ahn in 2012 (Hall, Ahn, & Greene, 2012). They draw a distinction between descriptive and prescriptive valuing. The former focuses on bringing out the values that different stakeholders have in relation to the evaluation at hand. The latter, prescriptive valuing, is more what you're saying: that evaluators can (and should) promote certain values through evaluation; for you, those values are grounded in a commitment to social justice.

I focus more on descriptive valuing and facilitating learning through sharing diverse perspectives on the projects being evaluated in my own practice. This is driven by the contexts that I work in for the most part. I'm usually subcontracted as an external evaluator for grant-funded projects. The expectation of the funders and the clients is that I am objective (I prefer the term "neutral") and independent. Within these constraints, I can argue for including a range of perspectives on and experiences with the evaluand through the data collected and shared to stimulate reflection and learning. I suspect my consulting practice would mostly evaporate if I advocated for values that could be seen as compromising my so-called objectivity. It must be nice to work in a field where there is a clear commitment to social justice. How does that play out in your work?

VALUING IN PRACTICE

SD: It definitely does make a difference because it is expected. For example, I attended the Shifting the Paradigm conference in 2017, which focused on HIV and Black women. This was the first conference of its kind; where Black women researchers, educators, practitioners, people living with HIV, including trans Black women, gathered to discuss HIV. Now there was definitely an elephant in the room because the larger conversation in HIV tends to be dominated by issues affecting gay White males and not necessarily those affecting Black and brown people who are disproportionately affected. Even in that room with Black identifying women, there was still tension between heterosexual Black women and trans Black women. This was an example of what inclusion should look like.

AK: How did the conference get to this inclusive space?

SD: In places like Atlanta where I live, doing subgroup-specific HIV outreach, the boots-on-the-ground approach, is critical. Now, mind you, not everybody in the room necessarily felt that way. There was an open forum at the conference, and some of the trans women took umbrage at something someone said. It got hot fast. Thankfully, we had an excellent moderator who was able to say, "Okay. This is an important issue, and we need to talk about it calmly." Because we were intentional about inclusivity, it was a space where hard conversations could happen.

AK: Listening to you, I'm struck by two things: the need to create a space for Black women affected by and working on HIV to come together and the need to challenge what it means to belong in that space. You were able to have a hard conversation about Black trans women and their place at the table. Not everyone wanted it, but as you say, that's probably the only place where such a conversation could happen.

SD: Yes, that is very true.

AK: You know, I work mostly in conventional evaluation spaces and don't often have an opportunity to engage a variety of stakeholders directly in the evaluation process. There are lots of reasons for this, but mostly it comes down to beliefs about evaluation as objective and distanced, particularly when contracted as an external evaluator. These beliefs come from outside the field of evaluation through funders, principal investigators, administrators, and policy makers—those groups that tend to get more voice in evaluation from the get-go. Recognizing these types of constraints, I find myself constantly maneuvering to make space for dialogue and learning.

I think back to one of the earliest projects I did of a school district seeking community input into its educational goals. We conducted 20-some-odd focus groups for the project with different segments of the community. The focus groups allowed for people to participate who usually weren't at this goal-defining table. Focus groups organized by role and affinity group enabled us to learn what different stakeholders had to say. Importantly, the focus groups illuminated the assumptions that groups had about other groups. I wrote up the findings by type of stakeholder (i.e., parent, teacher), hoping that the space afforded by reading about others and their experiences directly might make the room needed for listening and learning rather than reacting to expected stances as can happen in direct conversation. I shared the teacher focus group summary with the parents and the parents' summary with the teachers. Ultimately, I don't think there was time or interest on the district's side to host an opportunity for dialogue between the groups, but at least there was some exchange of perspective.

VALUING, INCLUSIVITY, AND SOCIAL JUSTICE

SD: There definitely can be a tension between a conceptual commitment to inclusive, social justice-oriented evaluation and what you are contracted to do, a tension heightened by context and constrained by awareness of what evaluation can be.

AK: Yup. Thinking about inclusion pragmatically, I spend time talking with the people whose program I'm doing an evaluation for, learning from them about the program, asking questions, suggesting ways that evaluation could be used, and encouraging them to think evaluatively. Really, a good bit of my evaluation work is educating people that I work for about what evaluation can do.

SD: This makes me think about values involved in judging quality in evaluation. Jennifer asked, "How will we know if it's good?"

AK: So, let me answer your question in an indirect way. I had the opportunity to look at that question of quality from a very meta-perspective a couple of years ago when I served as responsive evaluator for a project being run by Leslie Goodyear, another of Jennifer's students, and her colleagues. I call it my "meta meta-evaluation" project because it brought together about 25 individuals—including many seasoned evaluators—to form a STEM evaluation community of practice around evaluation capacity building to improve the quality of STEM evaluations.

Early on in the first convening, it became evident that numerous assumptions were embedded in that question of quality, derived in no small part by how individuals had been trained in and subsequently practiced evaluation. Quality could be defined narrowly as in the pronouncements that an evaluator makes about a project's accomplishment of its goals or broadly as in its contributions to making the world a fairer, more just place.

It was this project that really drove home to me the uniqueness of our training with Jennifer with its attention to values and how they manifest in evaluation practice. I feel that we benefited from that training in that we are willing to acknowledge that our evaluation choices are value-based, gave us language to talk about values and how they infuse practice, and with that knowledge, we are able to think more critically about how we do our work. For me, those values have to do with attending to voices that have a stake in decisions about quality and valuable perspectives that should be brought to bear on these decisions. How about you? What does judging quality look like in your work?

SD: For me, it is placing high value on inclusivity, and I think you cannot judge quality unless everybody's voice is heard. There was something I learned from the work of Asa Hilliard, who was an educator, researcher, and evaluator. Asa always said that context and history are always important. Are they present in the framing and the telling of the issue? More often than not, I find that they are absent in the work that I do, especially when it comes to HIV. When we look at most of the history of research, social science, and evaluation, it is usually the outsider's gaze that determines what is valued. What is defined as an issue? My work in Africana studies has helped me to figure out how to center research and evaluation in the lived experiences of the people, which, more often than not, causes me to do research on the history of an issue, thus highlighting the value of different stakeholders, especially those who are marginalized.

I use this example. I wrote an article in 2010 where I was looking at the issue of HIV and how it's typically an issue that is defined by public health using indicators based on people's behavior. How many contacts have you had? How many condoms? How many partners do you have? How many times do you use a condom? And those kinds of things. Actually, Asa Hilliard was the one who gave me a book called *Decolonizing Methodologies* by Linda Tuhiwai Smith.

In the book, Smith (2002) uses a symbol that looks like a bull's-eye. The outermost ring of the target represents survival, the next ring is recovery, followed by development, then self-determination at the center. Working from the outside to the center of an issue, we move from a position of survival to self-determination—thus, the questions change, and the values become more transparent. In the context of HIV, survival around prevention might be to pass out condoms. Recovery might involve training, and development might be where we start to develop curriculum. Self-determination raises other questions. When I apply this framework to HIV in the Black community, self-determination became the need to have more self-affirming conversations and programs and initiatives around positive self-love and the Black body. These types of questions go far beyond the typical public health approaches to HIV, right?

AK: Totally.

SD: Smith (2002) goes on to say once you identify what self-determination means, you have to unpack it. She offers four processes: healing, decolonization, transformation, and mobilization. What are the issues re-lated to "healing" around the Black woman's body? We can't deal with that question without thinking about enslavement and Jim and Jane Crow, the Black codes, rape laws, all of those things speak to how the Black female body has been commodified and treated in a particular way in the United States. What are the issues related to decolonization? How has the Black female body been colonized? We need to look no further than tropes about the Jezebel and Mammy. Then the question becomes, once we have identi-fied the history and the context, how do we then begin to transform around those issues and then eventually mobilize?

The issue of centering the question cannot be accomplished in a silo. Even the people who work in a given context may not know or value the history or context. So for me, and when you go through that process, you get a better sense of what it means to judge quality.

AK: Absolutely. What a gift Jennifer gave us in teaching that the kinds of questions we ask in evaluation are always imbued with values. So, too, are the judgments we render on quality, be it program or evaluation. They help some groups, and they hurt others. To not attend to the relationships between the questions that we're asking, the methodologies we use, the data we collect, and the potential impact of the findings reported, is to do evaluation uncritically. Context and history matter a lot.

Over the last few years, I've done some evaluations of programs designed to support students at risk for not completing high school. One evaluation was of a charter school that targeted students at risk of dropping out due to language, poverty, unstable home life, and other factors. This school was open to adolescents who had fallen behind in credits or were failing courses, as well as adults who had been out of school for anywhere from 3 to 20 years who wanted to earn their high school diplomas.

An objective of the evaluation was to identify alternative metrics for suc-cess for schools and students such as these, given that the usual ones—grade point average, graduation in 4 years, standardized test scores—were of limited value for understanding the lived experience of the students and what success meant in that context. One way we sought to identify these al-ternatives was through interviewing past students, both graduates and non-graduates, about their experiences and the meaning they held for them. We were particularly interested in hearing from the adults who had returned to high school rather than getting their GEDs because these are people who, as a group, we often don't hear from. Most adults we spoke with said they went back so they could get a better job or promotion. About as many, though, said that they returned to high school because they wanted to be role models for their own children. These adults talked about learning how valuable education was later in life as something they wanted their children to learn sooner than they had. It was good for us, as evaluators, to hear this as we sought to assess what success was in this context. It was good for the school administrators and the governing council to hear this because it helped them articulate important outcomes for older students that they may have sensed but lacked evidence to assert. And it was good for our state legislators to consider this information in their discussion of capping the age of participation in public education at 21 years old, legislators who may assume that a GED has equivalent value as a high school diploma without any other perspective to challenge the assumption's validity.

SD: As evaluators, we are contractors and typically outsiders to the context, which means we must be willing to do the work to educate ourselves on the background and context of the evaluand. These steps should improve our ability to say, "Yeah, this is good work." Hall, Ahn, and Greene underscore this sentiment in 2012 when they state that "evaluators must assume responsibility for explicating and justifying the values being advanced in their work, in ways that respect other values and thus other evaluation approaches and evaluators" (Hall et al., 2012, p. 206).

Last week in my graduate qualitative research methods class, we were discussing the phenomenon of Rashomon, a term from a 1950s Japanese movie where several people observe the same event, but everyone had a different interpretation of what was "true." For me, the interpretation of results is rooted in the cautionary tale of the 1965 Moynihan report that was commissioned under the Johnson administration as part of the Great Society efforts. They gathered data on the state of the Black family. The title of the report was *The Negro Family: A Case for National Action* (Moynihan, 1965). The report concluded that the Black family was in trouble due to the emasculating behavior of Black women undermining Black men, that he doesn't feel free to contribute to the family. Everything is centered around this interpretation without looking at the structural issues that kept the Black family apart.

While the Moynihan commission looked at the issues affecting the Black family, again about framing and the perspective, they looked at it from a deficit model and not a strength perspective, which is very common in social work. There was a whole body of literature on resilience that came out after that study was published. And so, the issue of interpreting results is, when you have a singular perspective guiding the interpretation, you overlook other worldviews. Some people would call it discriminatory observation.

Greene spoke to this issue in 2004 when she quoted from House and Howe (1999) saying,

> Evaluation always exists within some authority structure, some particular social system. It does not stand alone as simply a logic or methodology, free of time and space. And it is certainly not free of values or interest. Rather evaluation practices are firmly embedded in and inextricably tied to particular social and institutional structures and practices. (p. 122)

So let's come back around to values. What are the values you generally encounter in your work? How do you interact with them?

AK: It depends on the project and the people who are running the project, but mostly, I find that the clients I work for see evaluation in very instrumental ways: Did we do what we said? Have we proven that our model works that we can use for our next grant submission? Occasionally though, I will work with someone who is more interested in learning from evaluation. That's a joy even though it tends to be much more work. Once in a rare while, I have a client who wants to disrupt the status quo, but in my experience, that's been more research than evaluation. I don't blame folks for being narrow-minded about evaluation. That's what they know. That's how evaluation is presented in request for applications (RFA) and by grantors.

Going back to that distinction between descriptive and prescriptive valuing, I try to expand my clients' understandings of evaluation through how I do my work because I firmly believe that evaluation can be a place for learning from evidence and through dialogue. My prescriptive stance on evaluation is that it should be educative. I try to be agnostic with regard to descriptive values, foregrounding what people have to say for themselves and making room for multiple perspectives, not just the ones we usually hear. I often find myself saying to clients that I'm not going to make very good judgments about your projects if I don't really understand what's going on, with whom, and why, so we are going to have to engage with each other and talk, preferably on a regular basis. How do you engage values in your evaluation work?

SD: I take your approach as well because I don't want to be frustrated doing something that is not important to me, and I can't address it the way I think it needs to be. Much of that comes from acknowledging that a dominant pedagogy exists in academe. I teach my graduate research students that biases and values exist. There is no such thing as objectivity or value-free research or evaluation. Be reflexive so you know when your biases are showing up. So, what do we say about values?

AK: They're there.

SD: Yes, they're there.

AK: Don't pretend they are not. I really love what you said about acknowledging that there is a dominant way of doing things. If we recognize that there is a dominant way, it opens the possibility that there are alternative ways, and with that opened up, you can ask good questions about whether the dominant way is of value in this particular context.

SD: When we embrace a values-engaged approach to evaluation, we automatically open the front door to diversity, inclusivity, and pluralism. I've always seen evaluation as a tool for liberation. I kind of fancy myself as the Harriet Tubman of evaluation, and Harriet Tubman made this famous statement when she was asked how she was able to free so many people. Her response was, "I would have freed so many more if they had known they were enslaved" (https://professorbuzzkill.com/tubman-thousands-slaves-qnq). And this is the quandary. As evaluators, we have a commitment to leading people to their respective "Promised Lands." I see us as agents of change. We help people actualize their mission statements. I see evaluation as a Sankofa experience. Sankofa is a concept that essentially means you cannot plan your future without knowing your past. The image is that of a bird with one foot moving forward, but the bird is looking backward with the egg of her future in her beak. The literal translation is that you cannot move forward until you understand your past.

When we talk about holding space, it is really allowing all of those things to come together. The client has to trust us. We do that by focusing on democratization and bringing them under the umbrella that allows us to hold space for each other. Ultimately, we got that from our experience with Jennifer. She gave us the kind of methodological and theoretical freedom to operate in spaces we already knew we wanted to practice.

AK: Truly.

PARTING WORDS

In the same way that evaluation can and has been practiced conventionally to maintain the status quo, it can be leveraged to disrupt it. Jennifer taught us that. She offered us a language and set up concepts to be able to talk about values, valuing, and evaluation that was inclusive by nature and critical by design. By helping to diversify the field of evaluation, Jennifer brought people who asked different kinds of questions of programs they/we evaluated and helped us understand that these questions were and are legitimate to ask in evaluation. We experienced how including people not usually at evaluation tables in our work helped us be better evaluators, more sensitive to context and history. Perhaps most importantly, Jennifer showed us that we could practice engaged evaluation with people and programs without jeopardizing, and quite possibly enhancing, the quality of our work.

NOTE

1. We were talking about the call for abstracts one evening and decided to write together using a dialog format. Guided by a few predefined questions, we talked for about an hour and a half about Jennifer, about our training, and about our evaluation practice. Feedback from the book editors gave us an opportunity to further reflect on the dialog we had had, to write, to talk, to reflect, and to write some more. This chapter is the culmination of these conversations.

REFERENCES

Greene, J. C. (2011). Evaluation, democracy, and social change.

Hall, J. N., Ahn, J., & Greene, J. C. (2012). Values engagement in evaluation: Ideas, illustrations, and implications. *American Journal of Evaluation, 33*(2), 195–207. https://doi.org/10.1177/1098214011422592

Howe, E., & House, K. (1999). *Values in evaluation and social research.* Sage Publications.

Moynihan, D. P. (1965). *The Negro family: The case for national action.* Office of Policy Planning and Research, United States Department of Labor.

Smith, L. (2002). *Decolonizing methodologies: Research and indigenous peoples.* Zed Books.

SECTION IV

THE LAST WORD

To conclude the book, Jennifer C. Greene offers her reflections. Her reflections are based on an interview the editors of this book conducted with her to explore her thoughts on how she considers herself a disruptor in the fields of evaluation and mixed-methods research, the value commitments advanced in her work, and her contributions to social science inquiry.

CHAPTER 16

INTERVIEW WITH JENNIFER C. GREENE

As co-editors of the book, we thought it fitting to give Jennifer C. Greene the last word. With this in mind, we decided to end the book with an interview with Jennifer. As part of the interview, we solicited her thoughts about her role as a disruptor in the fields of program evaluation and mixed-methods research. We also inquired about her commitments to democracy, diversity, and dialogue; her role as a mentor (particularly to women and people of color); as well as her familial and institutional influences. To end the interview, we asked about her vision for the future of mixed-methods and evaluation research and what she wanted people to take away from her scholarship. Overall, Jennifer's responses to our questions provide additional insights about her and how her methodological achievements were cultivated and advanced.

For clarity, we organized our discussion by topic, focusing on key aspects of the conversation. While we know Jennifer, our interview with her brought us closer to her. We know when you read the interview, you will feel closer to her and her scholarship as well.

Disrupting Program Evaluation and Mixed Methods Research for a More Just Society, pages 227–243
Copyright © 2023 by Information Age Publishing
www.infoagepub.com
227

ON BEING A DISRUPTOR

Jori Hall: As you may recall, the title of the book is *Disrupting Program Evaluation and Mixed-Methods Research for a More Just Society: The Contributions of Jennifer C. Greene.* And, as a friendly reminder, we invited the chapter contributors to describe how they thought you were a disrupter in the field and your contributions. So, to start Jennifer, I thought it would be good to get your perspective on what you consider to be your key contributions to evaluation and mixed methods.

Jennifer Greene: I love being called a disrupter! I don't know if I've ever been called a disrupter before, but certainly, at this point in my life, it's a compliment. And I take it as such.

I didn't aspire to be a disruptor, but I think being in the Academy gives you a safe place to disrupt.

I think academics are supposed to disrupt in some ways—to challenge existing mantras or existing ways of doing things. And I think you can be a successful academic without disrupting, but... I wear that label [disruptor] with pride. And I think I'm in very good company. I think there's more than one of us. It is our job!

Jori Hall: Great, thank you for that! And then, just as a follow-up, can you share with us some ways that you think you have disrupted or challenged existing traditional ways of conducting program evaluation or mixed methods, Jennifer.

Jennifer Greene: I think it was just the luck of the draw that there were revolutions happening ... you know, academic revolutions as I was going through graduate school and then in the early years of my career. You know in graduate school, I did not even know what qualitative inquiry was!

And I was just positioned to be able to jump on the train and try this [qualitative inquiry] out because it filled in so many blank spaces in traditional evaluation. Because traditional evaluation was oriented to the questions and the existing problems that the program was intended to address but in a very narrow way. And with the privilege of being in the academy and the qualitative revolution, followed by many, many others critical of social science, it was a godsend I think.

And so, I was just able to fill in a lot of gaps in the field of evaluation through different ways of asking questions, different methodologies.

And then, later on, [I thought] why should we be fighting about methodology? We can use all of our methods. It's not a battle. So later on, that was just another part of the piece. So, I didn't set out to be a revolutionary. Although I came of age in the late '60s, which was a revolutionary time, I was never on the front lines, but I was always there.

It was this idea of not doing social science or evaluation in the ways we've always done it. So, I was positioned well to pick that up—which many people did. It wasn't just me. I think my history kind of positions me to be able to say, "Hey why don't we do this?"; "Why don't we do that?"

And I just loved learning about qualitative methodologies. We didn't have any qualitative methodologies in graduate school, but they were there. And so, I just thought this was a whole new world.

And most importantly, you were endeavoring to focus on the primary people in a program, which are the program participants.

The other methodological traditions often did a survey or something, but we could tell we could gather stories. We could have conversations.

So, I think just the timing was right for me to jump on that. That way of thinking about social phenomena, and ride the train, for a long time.

Another little piece of that is, after some time, there were many methodological innovations and revolutions that came along: critical social science or having a critical viewpoint. Those appealed to me and I was in a safe place in the academy to be able to at least ask questions that were reflecting critical social science and so on. And then things kind of started spiraling in all directions.

My favorite person to think of in this context is Patti Lather. Every week she had a new method. I didn't do anything that she did. I just sincerely admired her willingness to take giant steps to critique existing ways of doing things and her cleverness and being able to do this. She's a good friend of mine. I admire her greatly.

I don't see the world or methodology as she does. I didn't see myself going that far. Evaluation is ultimately a practical enterprise. It needs to be. She offered various post critical and various kinds of philosophical ways of thinking about the world, which I thought were all important for academics to consider, but had little relevance in the real world. So, there were boundaries.

ON COMMITMENTS OF DEMOCRACY, DIVERSITY, AND DIALOGUE

Jori Hall: I'm going to ask a different but related question, Jennifer, and I want to set this up by telling you how we organized the book. The book is divided into three sections and the sections are based on commitments or values that we have observed in your work. We think, for example, that you promote or are committed to democracy, diversity, and dialogue. Those are the three sections in the book. And so, I just want to ask you about each one.

Commitments to Democracy

Jori Hall: So, in terms of democracy, how would you describe your commitment to democracy?

Jennifer Greene: Two things come to mind.

I think the question of "Who does evaluation serve?" You know. "Whose interests are being served in an evaluation study?"

First, I think that's a contextual question. And while you just don't always have the power or the ability to choose the audiences for the evaluation, we can always add one or two [questions] or design a study that looks not just to the funders and administrators but to the program, to the participants, the neighbors, and the local groups in that context, whoever they might be.

Doing some activist work in terms of equity. So, I think, to take an evaluation contract, you have to be willing to follow it. And very often, that means answering the questions of those already in power; but that doesn't mean you can't add something to it to also

address questions of the people who are the participants. They are the ones who are supposed to be served
by this [program]. And I see that as democratic.

And second, it's endeavoring to be inclusive. That
is, who gets to have a voice in this endeavor. And with
social justice, an evaluation of a program justly serves
all major constituents—to the degree it's possible. You
couldn't do that with everybody, but you can always
make an effort to do so.

So, I think it was turning from the powers that
be—the commissioners of evaluation, the funders, the
leaders, and so forth or the leaders of the program—to
a more inclusive lens through which to generate questions and gather data and get feedback on the success
of the program.

Commitments to Diversity and Equity

Jori Hall: Let's think about or talk about a little bit of
your commitment to diversity. Jennifer, what are your
thoughts about your commitment to diversity and
equity?
Jennifer Greene: It's not a lot different than the first one. You all
know only too well, I'm a strong believer that evalua-
tion is a value laden enterprise.

Why do they call it evaluation? I mean you all know
this as well as I do, evaluators through their work of
evaluating, always advance some values and not others.

And it seems to me that attention to democracy,
diversity, voice, and respect are critically important
dimensions of what I would consider to be a defensible
evaluation.

Now, I could never do it [evaluation] in the way that
I wanted to in terms of funding, time, and so forth. The
evaluation proposal that I would write with some of you
and others would not be the one that was expected, because it really didn't follow the party line. But programs
are developed by people—usually well-educated special-
ists in something.

And then designed and implemented often without
the voice of the people that they intend to serve. And I
think, evaluation has a responsibility to make sure that

people who are intended to be the participants in this program have a meaningful and consequential voice. That's what diversity means to me: that we don't stay with the same people in power.

Ayesha Boyce: What are your thoughts about diversity and equity within our field in terms of evaluation practitioners and theorists and how it relates to your own work? Your own mentorship? Because that is touched upon in the book as well.

Jennifer Greene: Let me collect my thoughts here. Within the United States, which is really the only country I know well, we have collectively taken little baby steps towards equity and fairness across all the peoples in our country. So, the context of most of my work is the United States. And, I think we've collectively taken baby steps.

I think in the evaluation community, we have taken slightly bigger steps, but not very big. At least in my view. I don't want to bad mouth anybody, but I think there's a lot of token attention to diversity and equity—especially in terms of race and creed. And it's perhaps not as powerful as it could be.

The associations focus on prioritizing inclusion and making sure that all people who want to be evaluated have a good chance to do so. It should be standard practice that experienced evaluators don't even have to think about it. That is, it's standard practice to make sure that the people who are the intended participants in a program are part of the planning. And the implementation and the review and the discussion are not with token members but actual numbers with a voice. And I don't think that is standard practice today in evaluation, but I could be wrong.

Commitments to Dialogue

Jori Hall: I wanted to briefly get your thoughts about the commitment to dialogue that you have. Your dialectic stance and mixed methods infuse the importance of dialogue, for example.

Jennifer Greene: I think a lot of the origins of these thoughts and ideas were Freirean. A kind of grand vision as the ideal. I think I came upon that somewhere mid-career. It's philosophy. So, it's not a toolkit but it's a vision. And, it seemed to me that the idea of dialogue made enormous sense as a framework for evaluation.

And so, it's not as if everything was dialogic, but I liked to see how we could have a dialogue with the teachers who were implementing a program or something. I thought for dialogue, let's bring the group together and just try to see what we can learn and what we're missing. I think it was aspirational to have a habit dialogic kind of presence; so, the participants in the program or even the people running the program can contribute to what we're planning to do and can have a say, can have a voice.

I think that's where it came from the Freirean vision. And I just tried to design a study that the first step is to have some conversations with people and see what's on their minds. You know, meet with the big brass, the superintendent or the principals and meet with the people and the participants in the program. I think it comes from that kind of vision: a dialogic way of being in the world.

Having dialogue at its best is almost impossible, but that kind of inclusion: let's hear from everybody. Let's make sure that we're going to come up with something that will be of assistance to everybody. So that's how I recall it.

ON MENTORSHIP

Ayesha Boyce: As you reflect on all of the evaluators you have mentored—whether formally and informally—what role does diversity and equity play into that mentorship?

Jennifer Greene: I don't think I could have done much of the work that I did without folks like you. Because to this day, I remain somewhat uninformed. And despite having children of color, I remain somewhat uninformed or not fully informed about what life is like for many people of color in our society today. So, working with all of you and many others . . . at least you have some access that I don't. You have lived experiences that I don't.

I need a partner to go with me to talk to people who are the intended participants of the program or to talk to the organization that has designed the program and find out what they're thinking is, and so forth. I don't have the insight knowledge. And I just don't want to pretend that I do.

Ayesha Boyce: I know that you weren't necessarily looking to mentor people from diverse backgrounds. I think we kind of probably sought you out. People like Jori, myself, Gabi Garcia, and even Lorna to some extent.

People of color have come to you, looking to you as a mentor. And, in fact, I would say that you were an excellent mentor for us. Making sure that we have found our way into our own professions in the field that has not always been the most hospitable for people of color. So, even though you're not a person of color, I would say part of your legacy is mentoring evaluators of color.

I don't know if you've had a chance to reflect on your own mentorship as an advisor, as a mentor, as someone who has mentored a fair amount of people of color. And they're still doing well. I don't think they have, like, horror stories in working with you, which I can say is not the case with everyone who has mentored people of color.

Jennifer Greene: That's a great question! I have to think about it a little bit. I think I was just thrilled and delighted and really honored to be asked to be a mentor for you. I mean I thought, well, maybe it's because there's no other people of color.

For some of the evaluations, it was very important that there be people of color on the evaluation team. And I felt that not only were you making your own unique contributions to the evaluation, but it was necessary to have members of the evaluation team—especially for programs servicing other members of color—I felt a little awkward. So, I don't know that I explicitly sought it out because it's not my choice, it would be the students' choice.

Rodney Hopson: Maybe you don't think about it, Jennifer, because the way they are describing it sounds like a group of students.

One period in your academic life was here at the University of Illinois. But you have contributed to a lot of us who came through you, such as my colleagues at Cornell and women colleagues that I'm still in contact with.

And then you wrote all of my letters of recommendation for a 20-year period to Duquesne, Mason, Toronto, and Massachusetts. I followed you and your [American Evaluation Association] presidency purposely. You've brought a lot of us in. You are foundational. I was hoping I wasn't going to get emotional, but you laid a foundation for us here at Illinois after you left in which a lot of women students and students of color are still benefiting from. I just don't know if you see that.

I mean you have been a godsend in a lot of ways. And I don't know if it has been your general disposition to just say, "Okay, I'm gonna work with whoever comes my way." It just so happens that a lot of us people of color have relied on you to help us as a catalyst. I don't know if you think about that.

Jennifer Greene: Well, I haven't thought about it quite that way. I mean this is a two-way shared process where I have benefited as much, or more even, having worked with people like you.

Programs that I evaluated over the years were intended to serve members of our society that aren't receiving all that they should. I've never thought about it this way, like strategic. Like, "Oh, I want to make sure that I collected a variety of students of color so they can help me better understand how that community is working." I never thought of it that way, but I realized that that was happening over and over again. That sounds strategic. I don't mean it that way. I just meant that it worked. It worked really well for me and I felt that.

There was no way I could legitimately do an evaluation of a program designed for people of color with no person of color on the team. I couldn't make a connection to that community that I don't know. I mean it deeper than that.

FAMILY INFLUENCES AND EXPERIENCES

Rodney Hopson: Maybe you don't stop to think about the influence of those you've touched, but we do.

You know, I was thinking about your relatives and how
they influenced you. Your familial influences, you know.
And it was in that NDE [New Directions in Evaluation]
issue 29. I don't remember, but I think it was that colleague at Utah putting together those summaries. Autobiographical summaries you wrote about your parents.

Jennifer Greene: I am not sure what you're referring to, but I'm
open for whatever conversation on that you want to have.

Rodney Hopson: I'm not reading it right now. But, you spoke about your
mom and dad and those influences, and being a child
of the 60s. And, I seem to recall, there was a social fabric of responsibility that you grew up in.

Jennifer Greene: So, I don't recall. I mean I'm not sure if it's relevant: my parents' influence on me. But, it was profound
really . . . and I just fell out of the gate, I had some very
strong commitments to values that have stayed with me
through today. I think I attribute that to my parents,
especially my mom.

Overall, I was blessed with very caring and thoughtful parents. My, you know, kind of classical middle-class
White parents. My dad was a banker, and he always encouraged all of his children to really dig in and do well
in school and get a good job. And my mom worked in
the social services field after we all went to college. The
material things in life didn't matter to my mother and
they really didn't matter to my dad, but my mother was
all about what kind of person are you. Like, where's the
kindness? What have you done for somebody else today?
Or, she would just say, let's sit down a little bit and talk
about the day. I think my parents deeply influenced me.
Some blend of my father's now what to call it . . . whatever
power he had and my mother's commitment to generosity, kindness, honesty, and all of those things. I think,
who I am does this go back a pretty long way.

Rodney Hopson: Beautiful. Thank you so much for sharing.

SCHOLARLY INFLUENCES AND EXPERIENCES

Cornell

Rodney Hopson: Were you influenced by the sort of company you kept
as a scholar, and the places, right? I'm really interested

in what unique influence did the Cornell experience or the Illinois experience bring to your academic path?

Jennifer Greene: Well, my colleagues at Cornell were Bill Trochim and Charles McClintock. Those are my evaluation colleagues there. And we got along just fine. We supported each other and were in a wonderful department! There was just a lot of forward-thinking people in the department and we all got along well. Charles, Bill, and I were very different, but I think we worked well together.

University of Illinois

Jennifer Greene: And my experience at Illinois, just as some of you know, the students were the ones who were taking charge. A lot of very smart students, like Leslie and Valerie Caracelli, . . . I have to get my mind clear to think of many others, but it was a family. Illinois was like a family group committed to shared values, not always exactly the same commitments, but a purpose and the sense of vision for what we were trying to accomplish. And a commitment to each other to do well by each other. If you signed up to do something, then the group could count on you to do it. And then we had a lot of fun too.

Rodney Hopson: Do you think about how your work changed as well? I don't know if you have thought about how your work evolved in different places. Like the unique place that Illinois was for your ideas, over a period of time.

Jennifer Greene: I think, I would like to stretch my memory back to Cornell and some of the projects.

I don't know that it was Illinois itself or just that I have matured as an evaluator, as a person who felt more competent in doing things collectively and not saying you're going to do it my way. But raising questions and inviting the whole team to contribute to how this evaluation should be conducted and making sure that we brought the teachers in at some point.

It could have been Illinois, but it was also I had put 20 years into the field already. So, this was a safe and nurturing place in which to keep going and keep moving to see if we can bring some of these commitments and ideas to fruition. And goodness, what a ride!

TAKEAWAYS: EVALUATION EPITAPH?

Jori Hall: A goal of the book is to say, "Here is Jennifer Greene's work. Here are some of her major contributions. Her commitments." But, I'm curious, what do you want people to take away from your work?

Rodney Hopson: Right, do you have an evaluation epitaph?

Jennifer Greene: I think I'll say something, but reserve the opportunity to think about it a little bit in a more refined way.

I think we've talked and you raised some of the concepts and ideas and ways of implementing those that are closest to my principles and my ideals for evaluation, that little bit of inclusion and then have dialogue and those kinds of ideas.

And, I envision an ideal evaluation is highly dialogic and what an evaluation would do. Visit the various constituencies. The teachers. The parents. The kids. Get an educational evaluation and some of the powers that be, and have a conversation with each one of them about what they think are the most important questions and what they like about the program.

So, for evaluation to be an inclusive and dialogic process, we would not come in with an agenda. We would say, let's talk about the program and so forth. My evaluation practice approach is trying to be inclusive,... getting very different sets of perspectives. I mean, I'm thinking of dialogue with a capital D.

We Are Guests. We Are Visitors

Jori Hall: What are some takeaways you want people to have from your work?

Jennifer Greene: I will allow myself some additional time to think about this, but the things that are coming to mind are the importance of respect. Treating everyone involved in a program or its evaluation with honor and respect.

We've done work together. I'm sure we've all said more than once that we're visitors to wherever we are. We're guests. We're visitors.

And we need to act in this context with deep respect for everybody there. And to try to be highly inclusive in our conversations and observations and so forth.

So, it's not walking in assuming you know very much. Having a dialogue would be ideal, but having a conversation . . . or those kinds of exchanges to try to assure people that the teachers, the parents, whoever they are, that we are visitors here. We are guests. We want to make sure that we ask them the right questions.

And right now, we're just learning about the program. We will come back and have another conversation. It's this kind of ongoing process of getting to know one another, getting to trust one another, bringing ideas forward and getting reactions. And I see it as a kind of a process of moving forward in planning and implementing evaluation.

And the idea of being a guest is extremely important. We need them—the teachers and the parents of the kids—to help us to get it right.

Ayesha Boyce: So, you've talked a little about what you see as your legacy in the field. This idea that we come in and we're guests, and the fact that we want to be democratic and bringing different voices and especially those that are least well heard and being inclusive to help facilitate, you know, social justice, but also just because, just democratizing. I'm just curious, when you think about your legacy to me like that's your legacy for what evaluation should be, right?

But do you have any thoughts about your legacy? Who, you have worked with or those who you have mentored? Or what you would want them to take away from their time working with you or their time?

All the contributors to the book have been profoundly impacted by you. I think it speaks volumes that, in the middle of a pandemic, we send out an email saying, "Hey, can you do this one more thing? We want to write this book all about Jennifer's work and pretty much everyone said, "Yes," and that speaks volumes to the impact that you've had on their lives and our lives professionally. I think that has taught us that the lines between professional and personal are very blurred. And so, my question is, do you have any thoughts about your legacy for those who have worked with you in terms of those who you have mentored or those who you have been peer mentors with?

Jennifer Greene: I feel enormous happiness. I don't know. Pride is not the right word. Happiness when I hear about the accomplishments of the people who I've worked with. Who I know best. I just have this . . . I don't know. This thing in your chest you get. Just like this big balloon of joy and happiness. I just take a little pride, but you know I'm not the only influence here. So, joy and happiness, I think.

I was skimming through a bunch of the [book] materials. Yeah and all of it. All of the children . . . you know, are making huge impacts on the field. Not all of my children, but many. I just, I don't . . . I mean, I take some credit, but it's not my credit. It's that you have done this. You have taken it places. Taking it further. Taking more risks. Used more inventive, alternative forms of representation. Whatever it might be. I could open the window and shout out all these wonderful evaluators making the world a better place!

I just feel enormous. Enormous joy is really the right word, that I have made an influence on the field in the form of I just setting out to do what I thought was important. All of you have picked up the mantle in your own way, which is the way it should be. And some of these, I never set out to change the field, I just set out to do what I thought was important. And it just gives me goosebumps and shivers to think about the number of people I have influenced carrying this forward in your own way. Because the field that I entered was male dominated and White people dominated. Traditional paradigms dominated doing experiments or surveys; that's what the field was. And now look at us now!

I've been a contributor, but not the revolutionary I might want it to be. But disruptive okay! I'll take that, a child of the 60s always wants to be disruptive!

So maybe that was an influence. Yeah, I was reading through some of the stuff they [contributors of the book] think. It just made me smile. God you can't see that my smile just got bigger and bigger and bigger and bigger! As I said, I was able to witness and review. All these amazing thoughtful reflections and commitments. I just keep on keeping on, and now you have your own students, of course. So yeah, this was the luck of the draw in my life. I didn't set out to do anything like this, just took the opportunity.

ON THE FUTURE OF EVALUATION AND MIXED METHODS

Jori Hall: Given your contributions and what's happening now in the world, where do you think evaluation and or mixed methods need to go in the future, from your perspective? What do we need to be doing now? In the future?

Jennifer Greene: That's it, that's a very hard question. Because the future is not set in stone. The future could be a lot of different things. In the pandemic alone gives us a very uncertain future. It's a little bit hard to anticipate. I guess, I would say, I could say, what I would hope for. Yeah, hope for is a continuation of all the things that you guys have done, and that your fellow collaborators at this point can keep on keeping on. And maybe there will be additional opportunities for including different stakeholder voices. You know that's something that is part of the commitment of this group for example. It's very important to hear what people say and what they do.

I don't feel I can come up with a future or a way forward. Things are so in disarray. I think my hope would be that the pathway all of you have taken and all the others who are involved will continue on and refine and take a little side trail over here because it looks like it could be interesting and useful and have an impact for this program and these people. At this time, it's not like we have to keep improving or moving, but just keep on keeping on. Because I'm deeply impressed with all that you do, and that there might be little, as I say, little pathways, little opportunities, to do, to try something new, or to do something different.

The most important thing is that we maintain the inclusion of all key stakeholders in the work that we do. There might be other ways to do that, but it's very important that we don't assume that we know what's going on. I think it's very important to be as inclusive as possible in the planning, implementation, and review of evaluation work.

And I, it's interesting. I don't know how to think about the future. I don't have a very long future. You guys have more, but the pandemic is, just kind of like, stopped everything.

ANYTHING ELSE?

Jori Hall: This is an opportunity for you to say or add anything that you feel needs to be shared that maybe we didn't ask you about or other thoughts that you just have or things you'd like to follow up on.

Jennifer Greene: Oh, one other thing. I want to add to the previous response, which I know that some of you have taken on yourselves, is the use of alternative representations. You know... this notion that use of the arts can communicate far more than a dry set of results from a questionnaire... I think that can be profoundly meaningful. So, I just want to add that to your packet.

Rodney Hopson: Love you Jennifer.

Jennifer Greene: Yeah! Thanks! I worry a little bit about the future of mixed-methods methodologies. I think, for many people it's like a programmed activity. Here's the way that you mix. You add in a couple of interviews or something. I'm not going to say this eloquently, but I think mixed methods and evaluation are separate things you know. Evaluation is a profit profession. It's a pursuit of something but mixed methods is a methodology.

But, like other methodologies, mixed methods advances some values and not others. Involved some voices but not others. And I keep running into people or activities that put very narrow boundaries around the possibilities of mixed methods. I would hope that you guys, your students, and your colleagues, those who want to use mixed methods, will give us some insights that we have not addressed and questions we never thought of before.

Sometimes, all you want to do is a very small mix. You want to use a survey and then do a few interviews. That's all you have resources for and that's all you have time for. But, I think the potential of mixed methods is to be inclusive, to bring to the surface multiple kinds of values, multiple kinds of ways of knowing, and multiple kinds of ways of being.

That's the vision I have. It's not always possible but that's the vision I have. And I would hope that with this crazy world, we don't lose that vision. That it remains part of the methodology of mixing. It is a kind of grand learning engine for generating some insights about important phenomena that you wouldn't have generated otherwise. So, I see a lot of just kind of technical uses of mixed methods and not so much the values, and the ways of being, ways of knowing. There's nothing wrong with technical. Although, we want to make sure not to lose the rest of it.

Jori Hall: Okay, well Jennifer I'm just again so appreciative of you taking time out of your day to chat with us about these things for the book, and capture your remarks, your perspectives, and all of that. And just again, Thank you so much.

Jennifer Greene: It's me that's thanking you... Well, this is nice. And I am delighted with it, embarrassed by it, of course, but delighted with it. And want to thank the three of you for all the work you've put into this already. I don't like to spotlight, as you might know. I thank you! This has been a pleasure and a real, real boost to my morale. I'm indebted to you for all these efforts. And thank you, thank you, thank you!

ABOUT THE EDITORS

Ayesha S. Boyce, PhD, is an evaluation teacher, scholar, and practitioner. She is an associate professor of educational leadership and innovation at Arizona State University. Before pursuing her PhD at the University of Illinois at Urbana–Champaign, she was an education research associate (evaluator) for the Arizona Department of Education. Her research focuses on attending to value stances and issues related to diversity, equity, inclusion, access, cultural responsiveness, and social justice within evaluation—especially multi-site, STEM, and contexts with historically marginalized populations. She also examines teaching, mentoring, and learning in evaluation. She has evaluated over 55 programs funded by the National Science Foundation (NSF), U.S. Department of Education, National Institutes of Health, and Spencer and Teagle foundations. She encourages students to develop a strong methodological foundation, conduct studies based on democratic principles, and promote equity, fairness, inclusivity, and diversity.

Jori N. Hall, PhD, professor in the qualitative research and evaluation methodologies program at the University of Georgia, is a multidisciplinary researcher and evaluator focused on social inequalities and the overall rigor of social science research. Specifically, her work addresses issues of evaluation and research methodology, cultural responsiveness, and the role of values and privilege within the fields of education and health. She has published numerous peer-reviewed works in scholarly venues; she has authored the book, *Focus Groups: Culturally Responsive Approaches for Qualitative Inquiry and Program Evaluation* (Myers Education Press, 2020); and was selected as

Disrupting Program Evaluation and Mixed Methods Research for a More Just Society, pages 245–246
Copyright © 2023 by Information Age Publishing
www.infoagepub.com
245

a Leaders of Equitable Evaluation and Diversity (LEEAD) fellow by The Annie E. Casey Foundation. She currently serves as an external evaluator and co-principal investigator for programs funded by the National Science Foundation and the Robert Wood Johnson Foundation. She is also the co-editor-in-chief of the *American Journal of Evaluation.*

Rodney Hopson, PhD, serves as professor of evaluation in the Department of Educational Psychology, College of Education, University of Illinois–Urbana Champaign. He received his PhD from the Curry School of Education, University of Virginia with major concentrations in educational evaluation, anthropology, and policy and sociolinguistics. Hopson's research interests raise questions that (a) analyze and address the differential and ethical impact of education and schooling through policies and politics that surround marginalized and underrepresented groups and (b) seek solutions to ethical and political issues that ameliorate the issues faced by these marginalized and underrepresented groups in national and international domains with a particular interest in learners and groups of African descent in the United States and abroad. This scholarship has resulted in 8 (co-edited or co-authored) book volumes, 3 (co-edited) special issue volumes, and over 30 (authored or co-authored) peer-reviewed articles in venues such as *Anthropology and Education Quarterly; Diaspora, Indigenous, and Minority Education; International Journal of Human Rights; Journal of Negro Education; New Directions for Evaluation; Race Ethnicity and Education; Review of Educational Research,* and *Urban Education.* In addition, Hopson has co-authored over 30 book chapters in venues such as *Handbook of Urban School Leadership, International Handbook on Evaluation, Handbook on Practical Program Evaluation, The Handbook of Social Research Ethics,* and *The International Handbook on Urban Education,* and other titles in national and international social science and educational research. He is the former 2012 president of the American Evaluation Association (AEA) and recipient of the Marcia Guttentag Early Career Award (2000), Robert Ingle Award for Service (2010), and Paul Lazarsfeld Evaluation Theory Award (2017)—all from AEA.

ABOUT THE CONTRIBUTORS

Tineke Abma is the executive director of Leyden Academy of Vitality and Aging and professor "Participation of Older People" at Leiden University Medical Centre, Department of Public Health & Primary Care in The Netherlands. For over 25 years, Tineke Abma has been researching themes closely related to the participation of clients and citizens, participatory research, arts-based methods, ethics, and diversity. Her work aims to improve the social inclusion and quality of life of people, especially those in marginalized positions. The last 2 years she has supervised the first national study into the value and impact of arts in long-term care for older people, and a study into the lived experiences of people with Parkinson and MS enjoying improvisation dance. In 2013, she received an Aspasia laureate from the Dutch Council for Scientific Research (NWO) and her work has also been awarded for its high societal impact. Abma is the author/editor of a number of books, including *Evaluation for a Caring Society* (Information Age Publishing, 2018) and *Participatory Research for Health and Social Well-Being* (Springer Nature, 2019).

Cherie M. Avent is an assistant professor of evaluation in the Department of Educational Psychology at the University of Illinois at Urbana–Champaign. She received her PhD in educational research, measurement, and evaluation from the University of North Carolina Greensboro. Cherie's research focuses on issues related to social justice, communication, and STEM education research and evaluation with particular attention on contexts serving underrepresented minorities. Her recent publications highlight lessons learned

Disrupting Program Evaluation and Mixed Methods Research for a More Just Society, pages 247–257
Copyright © 2023 by Information Age Publishing
www.infoagepub.com
247

from the utilization of social justice-oriented evaluation approaches in STEM contexts and the experiences of evaluators of color within the field.

Sharon Brisolara is an educator, writer, and chief executive of Inquiry That Matters. As an organizational coach and program consultant, she continues to draw on what she learned under the leadership, tutelage, and care of Dr. Jennifer Greene. Throughout her career as a program evaluator and educational administrator focused on equity, inclusion, community development, and student care, she has been grateful for the focus on social justice, witnessing, allowing, embodied knowing, and call to action that she received while studying in the 1990s at Cornell University. Dr. Greene provided steadfast support as she proposed using participatory evaluation, new at the time, as the center of her dissertation work with women microenterprise owners in Costa Rica. Jennifer supported her and the co-authors of this chapter to navigate the politics of elaborating the feminist evaluation model. She believes that she has paid forward with what she was gifted to students, community members, and program staff.

Jill Anne Chouinard is an associate professor in the School of Public Administration at the University of Victoria, British Columbia, Canada. Her main research interests are in culturally responsive approachesto research and evaluation, participatory research and evaluation, and evaluation and public policy. Much of her evaluation work has been conducted in culturally and socially diverse community settings, where she has extensive experience leading evaluations at the community level in the areas of education, social services, public health, and organizational learning and change. She positions evaluation as a catalyst for learning, collaboration, equity, social justice, and community change. She is currently the editor-in-chief of the *Canadian Journal of Program Evaluation* and a section editor (Ethics, Values, and Culture) for the *American Journal of Evaluation.*

Janie Copple is an assistant professor in the Department of Educational Policy Studies at Georgia State University. She earned her PhD in qualitative research and evaluation methodologies from the University of Georgia. Prior to PhD studies, Janie enjoyed a 15-year career in K–12 education as a teacher, curriculum facilitator, and district-level coordinator. Her research interests include adolescent and family experiences with puberty and sex education and the role of philanthropy in secondary and higher education. Janie has published works in *The Qualitative Report* and *Qualitative Inquiry.* She is also co-editor and contributing author of *Conservative Philanthropies and Organizations Shaping U.S. Educational Policy and Practice.*

Sarita Davis has conducted culturally responsive program evaluation in marginalized communities for almost 30 years. The context of her evaluation work includes, but is not limited to bio-medical education, environmental health, faith-based education, foster-care, public health, and HIV prevention education. Her most recent body of research focuses on HIV risk and sexual communication between mothers who have survived childhood sexual abuse (CSA) and their daughters. Sarita's theoretical emphasis is on Womanism and the intersectional effects of race, class, and gender on health outcomes. She has served as an NIH investigator on several grants including an HIV and religion grant and an HIV prison re-entry project. The former explored the obstacles and opportunities of using faith-based social services as a conduit for early intervention. The latter grant assessed the efficacy of HIV+ prisoners achieving stable health outcomes. As a result of her scholarship, she has developed a theoretical and applied praxis for conceptualizing, designing, implementing, and evaluating community-based research in distressed urban communities. Sarita has parlayed her scholarship into peer-reviewed manuscripts and book chapters on culturally responsive assessment in the fields of evaluation, education, social work, mental health/therapy, Africana studies, and public health. However, she is most proud of her ability to educate graduate students in the fields of social work and Africana studies on how to incorporate culturally responsive approaches into their methodological practice.

Elissa W. Frazier is a curriculum & instruction EdD doctoral candidate at Loyola University Chicago. Her research interests include school improvement and policy evaluation, culturally responsive instructional design, and technology integration. She has a breadth of K–12 experience having served as an instructional coach, reading interventionist, and high school English teacher in urban schools spanning 12 years. In her current role as a research associate at the Education Development Center (EDC), Elissa works within the School Improvement Portfolio evaluation team to help educators, policymakers, and program officers make evidence-informed decisions about education, equity, and policy initiatives.

Molly Galloway is a PhD candidate in education policy organization and leadership (EPOL) at the University of Illinois at Urbana–Champaign (UIUC), where she was first introduced to both evaluation theory and methods under Jennifer Greene's tutelage. Molly is also a proud graduate of Grambling State University, a historically Black university (HBCU) in Louisiana, where she obtained a BA in sociology and much more. Although, over the years, her research and professional interests have shifted, she believes her interests can be best categorized in four ways: (a) abolition, (b) fugitivity, (c) liberatory education, and (d) evaluation for liberation.

Emily F. Gates, PhD, is an assistant professor of evaluation in the Measurement, Evaluation, Statistics, and Assessment (MESA) Department, in the

Lynch School of Education and Human Development at Boston College. Her research examines how evaluation theory, practice, and methodology can address the complexity of social programs and system change initiatives. Her published works include *Evaluating and Valuing in Social Research* (co-authored with Thomas Schwandt, Guilford Press); co-editor of a special issue of *New Directions for Evaluation* on systems- and complexity-informed evaluation; and peer-reviewed articles and book chapters on systems thinking, values, and equity in evaluation. She has a decade of experience conducting evaluations of social, educational, and health initiatives primarily in the areas of science, technology, engineering, and mathematics (STEM) education; K–12 teacher professional development; and tobacco control and prevention. Prior to her current faculty position, she was an evaluation fellow at the U.S. Centers for Disease Control and Prevention. She has a PhD in educational psychology with a specialization in evaluation from the University of Illinois at Urbana–Champaign.

Melissa Rae Goodnight is an assistant professor of evaluation in the Department of Educational Psychology at the University of Illinois at Urbana–Champaign and affiliate faculty member of the Center for Culturally Responsive Evaluation and Assessment (CREA). Her scholarship explores cross-cultural translations and applications of social justice theories in educational monitoring and evaluation (M&E) and research. With an emphasis on the interests of underserved and historically marginalized communities in the United States and India, her work spans issues of representation, validity, participation, and self-determination. Melissa's publications include articles on (a) translating critical race theory for analyzing discrimination in India's school system and (b) developing a framework for conducting research on the evaluation influence of M&E in India. Her upcoming co-edited book with Dr. Rodney Hopson, *Transformative, Comparative, and Intersectional Possibilities of Ethnography and Evaluation* (Emerald Insight), examines the connected use of ethnography and evaluation in scholarship across the world that employs contemporary strategies for engaging people, collecting data, generating knowledge, and taking action to make a social impact. Melissa began working in health and education while abroad as a U.S. Peace Corps volunteer in Kingston, Jamaica. She holds a PhD in education from the University of California Los Angeles with an emphasis in comparative education.

Priya Goel La Londe is an assistant professor at the University of Hong Kong in the Faculty of Education. A former teacher and school leader, she holds cross-disciplinary training in sociology, early childhood education, and education leadership and policy. Dr. La Londe's research examines the relationships between school reform policies and professional culture. Currently, she is investigating how performance accountability policies influence the identities and practices of teachers and school leaders in the contexts of Shanghai and Hong Kong. Dr. La Londe's research is supported by the Hong Kong Research Grants Council and the China Confucius Institute, and she has assisted on projects funded by the Spencer and William T. Grant Foundations.

Wenjin Guo, EdD, is a clinical assistant professor in research methodology at Loyola University Chicago from the People's Republic of China, a bilingual in English and Mandarin. Her dissertation focuses on how to improve Chinese international students' learning experiences from a culturally responsive lens. Her research scholarship also involves bilingual education, culturally responsive teacher education, educational program evaluation, transformative teaching pedagogy, as well as I-Ching to empower the historically marginalized communities to tell their counter narratives and establish cross-cultural and interdisciplinary conversations between the East and the West.

Leanne Kallemeyn, PhD, is associate professor in research methodology at Loyola University Chicago's School of Education. She has been the principal investigator of multiple evaluation projects in the field of education, which she also uses as opportunities to innovate evaluation practice. Her scholarship has focused on how practitioners, particularly teachers and school administrators, use (and do not use) data and evidence in their daily routines. She uses this knowledge to build evaluation capacity in schools and nonprofit organizations. She has published several articles in journals, including the *American Journal of Evaluation, Evaluation and Program Planning*, and the *Canadian Journal of Program Evaluation*. Leanne teaches courses in program evaluation, qualitative methodology, and mixed methodology.

Alexis Kaminsky, has conducted program evaluations for 30 years more or less as an independent consultant and a part-time faculty member at the University of New Mexico. Most of her work has been in educational contexts, both in-school and out-of-school. Over the last dozen years, her evaluation practice has focused on broadening participation in STEM. She is committed to the educative functions of evaluation, learning from diverse perspectives, and challenging dominant narratives about who and what matters in practice, programs, and policies.

Melvin M. Mark is a professor of psychology at Penn State University. He has served as president of the American Evaluation Association (AEA) and as editor of the *American Journal of Evaluation*. He has published numerous papers and chapters, as well as several books, related to evaluation theory, methods, and practice. Among his awards is AEA's Lazarsfeld Award for Evaluation Theory.

Alison Mathie is currently an associate of the Coady International Institute at St. Francis Xavier University, Nova Scotia, Canada, having retired from full-time employment as its research director in 2016. She earned her PhD in program evaluation and planning at Cornell University under the guidance of Jennifer Greene in 1996. Resuming a career in international development, she then worked at the Coady Institute, as a facilitator/adult educator and researcher, collaborating on numerous action research partnerships around the world and documenting citizen-led initiatives. She has co-edited two books: *From Clients to Citizens: Communities Changing the Course of Their Own Development* (Practical Action Publishing, 2008), and *Citizen-Led Innovation for a New Economy* (Fernwood Publishing, 2015). She continues her interest in participatory research and mixed-methods approaches to evaluation in her ongoing work with the Self-Employed Women's Association in India.

Rafiqah Mustafaa, PhD, is the assistant director of evaluation at the Collaborative for Academic, Social, and Emotional Learning (CASEL). In this role, Rafiqah designs and conducts internal evaluation projects to help CASEL learn about and continuously improve its technical assistance to states, districts, and schools, and resources for educators, researchers, and policymakers. Before this role, Rafiqah served as a research associate, helping CASEL's partner districts to continuously improve systemic SEL implementation in their contexts. Before joining CASEL, Rafiqah designed and implemented program evaluations and evaluation capacity-building projects for Illinois-based foundations, nonprofits, and government programs as an evaluation coach and independent consultant. Rafiqah earned a PhD and EdM in educational policy studies, with an evaluation specialization from the University of Illinois, Urbana–Champaign, and BA degrees in political science and sociology from Pennsylvania State University.

Eleanor Ngerchelei Titiml is a research methodology (PhD) doctoral student at Loyola University Chicago from the Republic of Palau, an island nation in the Micronesian region of the Pacific. Titiml's research and professional interests include but are not limited to educational systems and evaluation, school policy, culturally responsive methodologies, cultural understandings of social justice, indigenous evaluation, and evaluation capacity building. She has experience in learning and working within areas of higher education, college-access programs, government, non-profits, and community organizations in the United States and her home in Palau. Today, Titiml serves as a graduate assistant in Loyola University Chicago's School of Education and a research associate at Planning, Implementation, Evaluation Org.

Tristi Nichols is a senior international evaluation expert who manages her own practice with notable clients including the United Nations and international non-governmental organizations. As her work has a global focus, Dr. Jennifer Greene helped her to sharpen existing strengths. Dr. Greene's inspiring guidance transformed Dr. Nichols' work to not only excel in cross-cultural outcomes measurement in different languages (including sign language) but also use innovative, hands-on, and adaptable approaches when designing, managing, and executing mixed-methods evaluations. Through Dr. Greene's demonstrated commitment to the evaluation field, and especially her consistent involvement in the American Evaluation Association and attendance to the annual conference, Dr. Tristi Nichols followed suit and subsequently had the opportunity to serve on the AEA board with Dr. Jennifer Greene.

Lorna Rivera serves as a staff user researcher at Algolia. Her work focuses on uncovering user needs, motivations, and behaviors to guide the development of its Search-as-a-Service platform. She aims to do this work in both methodologically innovative and socially responsible ways. Previously, Rivera served in a similar capacity at Twilio, Inc, a communication solutions company that powers our daily interactions. Rivera has a nearly 10-year history of academic work from her time at the Georgia Institute of Technology and the University of Illinois at Urbana Champaign. During this time she conducted evaluations primarily funded by the National Science Foundation's Division of Advanced Cyberinfrastructure to work with multiple high-performance computing centers and organizations around the world including Compute Canada, PRACE, RIKEN, Women in High-Performance Computing, and XSEDE. Rivera received both her Bachelor of Science in health education and her Master of Science in health education and behavior from the University of Florida. Prior to joining Twilio, Rivera worked with various institutions, including the Georgia Institute of Technology, University of Illinois at Urbana–Champaign, the March of Dimes, Shands HealthCare, and the University of Florida College of Medicine.

Thomas A. Schwandt, PhD is professor emeritus, University of Illinois, Urbana-Champaign; editor emeritus of the *American Journal of Evaluation*; member, editorial board of *Evaluation: The International Journal of Theory, Research & Practice*. In recent years his research has focused on the role of values in evaluation and social research, as well as on evidence-based reasoning and data-driven decision making. His most recent book is co-authored with Emily Gates, *Evaluating and Valuing in Social Research* (Guilford Press, 2021). Other publications include *Evaluation Foundations Revisited: Cultivating a Life of the Mind for Practice* (Stanford University Press, 2015); *Evaluation Practice Reconsidered* (Peter Lang, 2002); with Edward Halpern, *Linking Auditing and Meta-evaluation* (Sage, 1988); and, with Ken Prewitt and Miron Straf, *Using Science as Evidence in Public Policy* (National Academies Press, 2012). He has co-edited *Exploring Evaluator Role and Identity* (with K. Ryan, Information Age Press, 2002) and *Evaluating Educational Reforms: Scandinavian Perspectives* (with P. Haug, Information Age Press, 2003). In 2002, he received the Paul F. Lazarsfeld Award from the American Evaluation Association for his contributions to evaluation theory. He currently works as an independent evaluation advisor.

Denise Seigart is a registered nurse and dean of the College of Health Sciences at East Stroudsburg University, PA. She began her career as a registered nurse in 1980 and pursued her doctoral work with Dr. Jennifer Greene in the 1990s while teaching nursing full-time and raising two small children with her husband Bill. Denise was able to complete her doctoral degree and pursue a rich and rewarding career because of the mentorship of Dr. Jennifer Greene. She served as a faculty member and interim associate provost at Mansfield University, an assistant director of nursing at Boise State University, and as mentioned, now serves as dean at East Stroudsburg University. Dr. Seigart utilizes all that she learned from Dr. Greene in her work every day; conducting program evaluations, maintaining accreditations, and fostering community development and growth.

Kathryn A. Sielbeck-Mathes is president, CEO of Measurement Matters, Personalized Evaluation Services, LLC. Her first introduction to the field of program evaluation was in the summer of 1993 when she took a qualitative methods course at Cornell University, taught by Dr. Jennifer Greene. Dr. Greene's incredible passion for using evaluation to make a difference among vulnerable populations ignited a flame for the same in Dr. Mathes' career and subsequently changed its trajectory from nursing education and a nurse practitioner role to practicing as a full-time professional program evaluator. Upon graduating with a PhD in program planning and evaluation she founded her own evaluation consulting business in the northeast, then worked as a senior program evaluator, a director of Evaluation, and a vice president of Research and Evaluation for a large mid-Western research institute. She has recently re-opened her consulting business. Dr. Mathes attributes her responsiveness to diversity, attention to voices of the most disadvantaged, inclusiveness, and critical reflection to Dr. Greene's lectures, her publications, her presentations, and most importantly her example as an evaluator who believes in social justice for all, a place at the table for those not seen or heard, the importance of incorporating participatory and mixed methods into evaluation designs, and all while attending to rigor and the production of high-quality evaluative information.

Carolina Hidalgo Standen is an assistant professor in the Department of Education at the University of La Frontera, Temuco–Chile. Between 2011 and 2015, as a Fulbright scholar, Carolina developed her doctoral studies at the Department of Educational Psychology at the University of Illinois in Champaign–Urbana, where she conducted a study on teacher evaluation in Chile, under the guidance of Dr. Jennifer C. Greene. Currently, Hidalgo-Standen teaches undergraduate and postgraduate courses in the teacher education program, focusing on research methods in education, mixed methods, and qualitative research. She directs a research center on applied pedagogical research, whose mission is strengthening the school system, through the production and transfer of applied knowledge on school management, teaching-learning processes, and teacher professional development. In addition, Professor Hidalgo participates in various research training programs for teachers currently working in the school system. In her research work, Dr. Hidalgo Standen analyzes the sociocultural meaning of the concept of teacher quality, and its relationship with teacher education and teacher's professional development policies in Chile. Its contribution lies in questioning the theoretical and methodological assumptions underlying the implementation of the National Teacher Evaluation System in Chile, proposing an evaluation model that takes into account the different values associated with teaching work and that dialogues with the diverse socio-cultural contexts where teachers develop their teaching practice. In addition, she analyzes the cultural relevance of teaching practices in rural contexts, based on the study of differences in the way of learning of Mapuche children, belonging to the largest indigenous group in Chile—the Mapuche people.

Rebecca M. Teasdale, PhD, assistant professor of educational psychology at the University of Illinois at Chicago (UIC), is an evaluation methodologist who examines the valuing process in evaluation. She investigates questions about which values—and whose values—are privileged in the criteria that underpin evaluations. She is particularly interested in the specification of criteria in social justice-oriented evaluations, such as those informed by culturally responsive and democratic evaluation approaches. Dr. Teasdale develops methods for representing community and program participants' values in the specification and application of evaluative criteria. She applies these methods in a range of educational contexts, including adult and family literacy; informal science, technology, engineering, and mathematics (STEM) education; and special education. Dr. Teasdale currently designs and teaches graduate-level courses in evaluation methods, evaluation theory, and qualitative research methods. Prior to joining the UIC faculty, she served as a professional evaluator focusing on informal and formal STEM education, a science and mathematics librarian, and a molecular biology research associate. She holds a PhD in educational psychology with a specialization in evaluation from the University of Illinois at Urbana–Champaign

Veronica G. Thomas, PhD, is a professor in the Department of Human Development and Psychoeducational Studies at Howard University. She also serves as the evaluation and continuous improvement (ECI) director for the Georgetown–Howard Universities Center for Clinical and Translational Science (GHUCCTS). Her research interests include culturally responsive evaluation; physical and psychological well-being of Black families, with particular emphasis on women and girls; and the academic and professional development of students of color. She is the lead author of the recently published text, *Evaluation in Today's World: Respecting Diversity, Improving Quality, and Promoting Usability* (SAGE Publications, 2021). Over the years, Dr. Thomas has published work in numerous refereed journals including the *American Journal of Evaluation, New Directions for Evaluation, Journal of Community Genetics, Journal of Black Psychology, International Journal for the Advancement of Counselling* (published in the Netherlands), *Family Relations, Adolescence, Educational Leadership, Journal of Adult Development, Review of Research in Education, Journal of Negro Education, Sex Roles, Journal of Social Psychology, Women and Health*, and the *Journal of the National Medical Association*. Her work has been funded by the National Science Foundation, National Institutes of Health, U.S. Department of Education, and Women's College Coalition. Dr. Thomas's major professional associations include the American Evaluation Association (AEA), American Psychological Association (APA), and American Educational Research Association (AERA). In 2019, she received the AEA Multiethnic Issues in Evaluation Scholarly Leader Award for scholarship that has contributed to social justice-oriented, equity-focused, and/or culturally responsive literature.

Julian Williams, PhD is an evaluation officer at the MacArthur Foundation. In this role he supports all aspects of the Foundation's evaluation activities. In collaboration with program teams, leadership, and external partners, he facilitates evaluation design, implementation, and reporting. Before joining the Foundation, Julian was senior partnerships manager at the Partnership for College Completion (PCC), where he led a team that worked with Illinois' colleges and universities to eliminate gaps in degree completion by race and socioeconomic status through equity-centered programming and coaching for administrators, staff, and faculty. Prior to PCC, he was CEO and research evaluation consultant at Institution Builders, Inc., an independent consulting company that worked with nonprofits, government agencies, school districts, and foundations to improve important aspects of their work. Julian has a PhD and MEd in educational studies with a concentration in evaluation from the University of Illinois, Urbana–Champaign as well as a BA in educational studies and in English literature from Denison University. Julian is a proud Chicago public schools graduate, a Posse Scholar, a Diversifying Faculty in Illinois Fellow, and a Chicago Surge Fellow.

Julian is an advisory board member of the Pathways Initiative, a collaborative to strengthen the diversity of the Chicago area evaluation field by increasing the recruitment, training, and retention of culturally responsive and equity-focused evaluators of color in the Chicago region.

Susan Woelders is a researcher at Leyden Academy on Vitality and Aging and Leiden Medical University (LUMC) in The Netherlands. She has a background in organizational anthropology and conducted her thesis on patient participation in health care and research settings. She was involved in many participatory research projects in different sectors. In her work, she focuses on bringing together the knowledge and perspectives of all involved, in order to facilitate a mutual learning process and come to new insights. In her research, she aims to open spaces for the perspectives and knowledge of all involved, especially people in marginalized positions.